The Dynamic Properties of Supercooled Liquids

The Dynamic Properties of Supercooled Liquids

GILROY HARRISON
*Department of Electronics
and Electrical Engineering
University of Glasgow*

1976

ACADEMIC PRESS
London   New York   San Francisco
A Subsidiary of Harcourt Brace Jovanovich, Publishers

ACADEMIC PRESS INC. (LONDON) LTD.
24/28 Oval Road
London NW1

United States Edition published by
ACADEMIC PRESS INC.
111 Fifth Avenue
New York, New York 10003

Copyright © 1976 by
ACADEMIC PRESS (INC.) LONDON LTD.

All Rights Reserved
No part of this book may be reproduced in any form by photostat, microfilm,
or any other means, without written permission from the publishers

Library of Congress Catalog Card Number: 75-39966
ISBN: 0-12-328150-4

Printed in Great Britain by
Whitstable Litho, Straker Brothers Ltd.

Preface

This book describes the information which may be obtained from a study of the viscoelastic, ultrasonic and dielectric behaviour of supercooled liquids. The material was originally presented at a one-term seminar on the "Mechanical Properties of Liquids" given in the Department of Mechanical Engineering and Astronautical Sciences, Northwestern University, Evanston, Illinois. The collection of the widely scattered material into book form, with the advantage of a uniform system of units and notation, resulted from the interest of the audience and the enthusiasm and encouragement of Ralph Burton, who made possible my visit to Northwestern University.

I should like to express my gratitude to John Lamb for the help and guidance he has given me over many years, and to A. J. Barlow, W. W. Graessley and C. D. W. Wilkinson, among many others, for their critical reviews of parts of the manuscript. I am grateful to Judy Kozlov and Anna Cunningham for their expert typing of the manuscript.

Glasgow, G.H.
December, 1975

Acknowledgements

I am indebted to the following organisations for permission to reproduce illustrations: American Institute of Physics: Figures 2.3, 4.6, 4.15, 4.16, 5.14, 5.24, 6.16, 6.17, 6.18 and 6.19. American Society of Mechanical Engineers: Figure 2.5. Chemical Society: Figures 5.19, 6.23 and 6.24. Institute of Physics: Figures 4.5 and 4.8. Institution of Mechanical Engineers: Figure 5.3. John Wiley: Figure 2.4. Nature: Figure 5.23. The Royal Society: Figures 5.1, 5.2, 5.4, 5.5, 5.6, 5.7, 5.8, 5.9, 5.10, 5.11, 5.12, 5.17, 5.18, 5.20, 5.21, 5.22, 5.25 and 5.26. The Society of Rheology: Figure 4.10.

Contents

Preface v

1. Introduction 1

2. The Liquid State 11
   2.1. Density 12
   2.2. Viscosity 15
      2.2.1. Viscosity units 17
      2.2.2. Viscosity values 18
      2.2.3. Non-Newtonian behaviour 19
      2.2.4. Variation of viscosity with temperature . 20
      2.2.5. Variation of viscosity with pressure . . 28

3. Linear Viscoelasticity 31
   3.1. Propagation of plane waves in a liquid . . . 32
   3.2. The Maxwell element 39
      3.2.1. Oscillatory response of the Maxwell element . 40
      3.2.2. Transient response of the Maxwell element . 41
   3.3. The Voigt element 43
      3.3.1. Oscillatory response of the Voigt element . 45
      3.3.2. Transient response of the Voigt element . 46
   3.4. Viscoelastic models for complex systems . . 46
      3.4.1. The relaxation spectrum 52
      3.4.2. The retardation spectrum 53
   3.5. Interrelations between viscoelastic functions . . 55
   3.6. Method of reduced variables 56

4. High Frequency Methods for Measuring the Mechanical Properties of Liquids 63
   4.1. Reflection coefficient techniques 65
       4.1.1. Reflection of a plane shear wave at an interface, for normal incidence 66
       4.1.2. Normal incidence pulse technique . . . 69
       4.1.3. Inclined incidence technique 72
       4.1.4. Electrical system 73
       4.1.5. Acoustic system 76
       4.1.6. Very high frequency techniques . . . 80
       4.1.7. Pulse superposition methods 81
   4.2. Travelling wave techniques 84
       4.2.1. Torsional wave technique 84
       4.2.2. Travelling shear wave technique . . . 88
   4.3. Resonance technique 90
   4.4. Longitudinal wave techniques 91
   4.5. Optical methods 96
       4.5.1. Optical diffraction 96
       4.5.2. Brillouin scattering 100
       4.5.3. Stimulated Brillouin scattering . . . 106

5. The Viscoelastic Properties of Supercooled Liquids . . 109
   5.1. The high frequency limiting elastic modulus, G_∞ 110
   5.2. Viscoelastic relaxation behaviour of supercooled liquids   116
   5.3. Viscoelastic retardation in supercooled liquids . . 132
   5.4. Viscoelastic behaviour of liquids at high pressures . 141
   5.5. Theories of viscoelastic behaviour 145

6. Structural and Dielectric Relaxation 151
   6.1. Structural relaxation 151
   6.2. Propagation of longitudinal waves in liquids . . 153
   6.3. Analysis of structural relaxation data . . . 156
   6.4. Structural relaxation in supercooled liquids . . 163
   6.5. Dielectric properties of supercooled liquids . . 173

References 187

Index 197

1. Introduction

This book is concerned with the dynamic behaviour of liquids when the existing molecular equilibrium is disturbed by an applied stress. The stress may be mechanical, either shear or compression, or in the case of polar molecules it may be applied electrically. The attainment of a new equilibrium state following the stress application is not instantaneous, but takes a finite time which depends on the ability of molecules to move relative to their neighbours. The observed response of the liquid depends on the relative duration of the applied stress compared with the time constant, or relaxation time, which is associated with the change in equilibrium.

If the time over which the stress is applied is long compared with the relaxation time then the liquid can respond to the stress and will take up a new configuration. The liquid properties show no time-dependent behaviour and have the usual steady state values. Conversely, if the period of the stress application is short compared with the relaxation time the material cannot respond before the stress is removed, and the existing equilibrium state is unchanged. The liquid properties then reflect a situation in which molecular motion does not take place and the existing molecular arrangement is effectively "frozen". In this situation the response of a liquid to a shear stress, for example, is elastic deformation instead of the more usual viscous flow. Measurements made in between these two extremes of time scale, in the "relaxation region", enable the relaxation time and the rate constants of the equilibrium to be determined.

In principle, any form of time dependent stress function may be used and the liquid properties evaluated from the ratio of the resulting strain to applied stress. For practical reasons, and for ease of analysis, only two types of stress variation are used, either a step function or a harmonic variation at a specified frequency. A study of the transient response to an applied step of stress, or strain, is only convenient when the times involved are of the order of a second or longer. For shorter times the use of a sinusoidal stress variation is preferable, and the liquid parameters are studied as a function of frequency. These two approaches are complementary, and data obtained as a function of frequency can be readily transformed into the time domain, and vice-versa.

When subjected to simple plane shear the normal response associated with a liquid is viscous flow. The elastic response is only apparent when the time scale of the deformation is comparable with the time taken for molecules to diffuse to a new equilibrium position. In the region where the response is changing from purely viscous to purely elastic both effects are present and the liquid exhibits viscoelastic behaviour. Two forms of transient experiment can be carried out, stress relaxation and creep. In a stress relaxation experiment, a sudden strain is applied to the liquid and the deformation is then held constant. The decay of the stress is then measured as a function of time as the new equilibrium position is reached. In the converse experiment, creep, the deformation of the liquid following the sudden application of a defined stress is measured. Alternatively the recovery of the material following the sudden removal of the stress can be observed (creep recovery). These techniques are valuable when the viscoelastic relaxation times are of the order of a second or longer. This situation only occurs near the glass transition temperature T_g when the viscosity is of the order of 10^{12} Pa s. At temperatures above T_g where the viscosity is considerably less the relaxation time is correspondingly shorter: to obtain a period of stress application which is comparable to the relaxation time it is necessary to apply a cyclic stress variation in the form of an elastic wave. Two types of wave propagation are possible in isotropic media, shear waves and longitudinal waves. A plane longitudinal wave contains both a pure shear component and a compressional component, and the response to such a wave depends on both the shear modulus and the bulk modulus of the liquid. A plane shear wave introduces no volume changes and so in this case the behaviour is related only to the shear modulus. Thus in order to investigate effects which depend on the bulk modulus it is necessary to determine the shear behaviour first and then remove the shear contribution from the response to the longitudinal wave. Because longitudinal waves are sound waves at frequencies above the audible range, the study of material properties using such waves is usually referred to as "ultrasonics".

The compressional component in a longitudinal wave results in a cyclic adiabatic volume change, and the local temperature will vary in phase with the changing volume. This variation will perturb any equilibria which are temperature sensitive, such as rotational isomerism, energy transfer between translational, vibrational and rotational degrees of freedom and various types of chemical reaction. Such temperature induced effects are not considered in this book, but are adequately dealt with in many other texts, for example Herzfeld and Litovitz (1959), Gooberman (1969) and Bhatia (1967). The effects discussed in this book are those which depend on molecular flow between the regions of high and low density which are produced by the cyclic pressure variation associated with the longitudinal wave. The response can be charactcrised in terms of a volume, or structural, relaxation time, which is the time required for a change to a new equilibrium volume following a sudden pressure change.

If a liquid contains polar molecules, the equilibrium orientation of these molecules can be perturbed by an electric field. Such a change in orientation can be characterised by a dielectric relaxation time and produces a complex permittivity which is frequency dependent. The phenomena of viscoelastic, structural and dielectric relaxation thus all reflect the relative motions and reorientation of molecules with respect to their neighbours and it is therefore natural to suppose that the effects are related. The purpose of this book is to review the current state of knowledge of these phenomena, and the correlations which exist between them, as determined from measurements of the dynamic behaviour of liquids.

The most obvious liquid property which depends on the ease with which molecules can move relative to each other is the viscosity. While there are important differences between viscoelastic, structural and dielectric relaxation effects, in all three the relaxation times have a dependence on temperature which is similar to that of the viscosity. A useful measure of the order of magnitude of the relaxation times is the Maxwell relaxation time $\tau_m = \eta/G_\infty$. In this expression η is the viscosity, and G_∞ is the limiting value of the shear elastic modulus at high frequencies, or short times. The value of G_∞ is found to be close to 10^9 Pa for most liquids, a value which is typical of the glassy state, and it varies inversely with temperature. In contrast the viscosity tends to vary exponentially with temperature, and can change by many orders of magnitude over a small temperature range. Consequently, τ_m has a temperature dependence which is dominated by the viscosity, the value being given approximately by $\tau_m = \eta \times 10^{-9}$ s, where the viscosity η is measured in Pa s (= 10 Poise).

The highest frequency at which elastic waves can be readily generated is around 10^9 Hz, corresponding to a time period of 10^{-9} s. It is therefore not possible to use elastic wave techniques to observe relaxation phenomena in liquids which have viscosities less than 1 Pa s. The majority of simple organic liquids have room temperature viscosities considerably less than this value, and can therefore only be studied if the viscosity can be substantially increased, either by cooling, or by the application of hydrostatic pressure. It is usually easier experimentally to increase the viscosity by cooling than by applying pressure, but this can only be done in liquids which do not crystallise. Most simple organic liquids have a viscosity at the melting point of 0.01 Pa s or less. High values of viscosity can therefore only be reached in liquids which can be supercooled and remain in the liquid state at temperatures well below the melting point. The lower temperature limit of the liquid state is the glass transition temperature T_g, which may be conveniently identified as the temperature at which the viscosity is 10^{12} Pa s. The relaxation time is then 10^3 s, and changes in the molecular arrangement cannot take place during the time scale of a conventional experiment: the material can then be regarded as a glass.

It follows from the above argument that high viscosity materials, such as high molecular weight polymers, can be investigated using techniques operating at low frequencies, say below 10^3 Hz, and by creep and stress relaxation experiments. These techniques, and the viscoelastic behaviour of polymers are well documented (for example, Ferry, 1970, Eirich, 1958) and are not discussed in this book. The study of low viscosity liquids has only been made possible by the development of techniques operating at frequencies above

10^4 Hz, the pioneering work being carried out by W. P. Mason, H. J. McSkimin and colleagues at the Bell Telephone Laboratories in the late 1950's. More recently advances in techniques using light scattering have enabled measurements to be obtained in the frequency range 1–10 GHz, using the elastic waves which occur naturally in liquids.

The aim of this book is to survey the current state of knowledge of the shear, structural and dielectric relaxation behaviour of simple (i.e. non-polymer) supercooled liquids, and to describe the techniques by which experimental data are obtained, thereby giving an understanding of the limitations and the reliability of the available data. A knowledge of the dynamic properties of liquids is of both fundamental and practical importance. When compared with the solid and gas phases, the liquid state is poorly understood, and a considerable increase in the amount of experimental data is required to support and test theoretical developments. Existing theories of the liquid state fall into two main groups. One group consists of theories which attempt to evaluate macroscopic liquid properties from a rigorous analysis of the interactions between molecules. Solutions to these analyses are only valid for the simplest of molecules, such as the liquified rare gases xenon, krypton and argon. It is possible to calculate the viscosity of these simple liquids and to estimate the order of magnitude of the limiting modulus G_∞ and its dependence on pressure and temperature. These theories can not be tested however, as the low viscosity of the liquids makes the experimental determination of G_∞ impossible with current techniques. In liquids containing more complex molecules, for which experimental data are available, the assumptions made in obtaining a solution are no longer valid. Introductory reviews of these statistical mechanical treatments are given by Driesbach (1966), Pryde (1966) and Rowlinson (1969). An alternative approach is to formulate empirical hypotheses which will describe the experimental results. These range from simple empirical equations, e.g., Davidson-Cole and Barlow, Erginsav and Lamb (hereafter referred to as B.E.L.), which adequately describe a limited range of observed behaviour, to hypotheses which postulate a particular type of molecular motion, in many cases a diffusion process, and then attempt to deduce the consequent macroscopic behaviour. This latter approach, which is discussed briefly in Chapter 5, suffers from the inherent defect that the parameters involved cannot be determined from fundamental molecular or thermodynamic data, and therefore cannot be used to predict the behaviour of liquids for which experimental results are not available.

These empirical theories have considerable value, however, in describing liquid behaviour in many practical situations. In particular, information about the manner in which liquids respond to stress, in both shear and compression, is of direct relevance to an understanding of the processes occurring in lubrication. At the present time the behaviour at the high hydrostatic pressures which are encountered in elastohydrodynamic lubrication is of considerable interest.

Elastohydrodynamic lubrication occurs in heavily loaded rolling contacts, such as between gear teeth and in roller bearings, when the elastic deformation of the contact surfaces is significant as compared with the thickness of the lubricant film. A general account of elastohydrodynamic lubrication is given by Dowson and Higginson (1966). The theoretical analysis is well established in general, but several anomalies remain. The usual method of study is with a rolling-contact disc machine. Two disc-shaped rollers, which have their axes parallel, are pressed into contact, the lubricant forming a film between the rollers at the point of contact. Measurement of the tangential force at the periphery of the disc enables the mean viscosity of the lubricant to be determined for different conditions of rolling speed (the mean peripheral speed of the discs) and sliding speed (the difference in peripheral speeds). The results of this type of experiment (for example, Smith (1960), Crook (1963) and several later workers) give rise to several unexplained effects. In particular, the mean viscosity of the lubricant, measured at low sliding speeds, is observed to decrease with increasing rolling speed. When this data is extrapolated to zero rolling speed, the mean viscosity can be plotted as a function of the peak pressure in the contact. The mean viscosity is then found to increase in a roughly exponential manner with pressure up to a pressure of about 0.7 GPa, in accordance with conventional viscosity measurements. At higher pressures, however, a break in the curve occurs and thereafter the mean viscosity increases much less rapidly. This change in behaviour implies that either some change in the equilibrium fluid properties occurs at high pressures, or that the fluid is exhibiting viscoelastic behaviour, and the elastic and time dependent properties become significant. At increased sliding speeds the behaviour becomes dominated by heating effects and by non-linearities in the relation between shear stress and shear rate.

Crook (1963) and Dyson (1965, 1970) have attempted to explain these effects by considering the lubricant as a viscoelastic fluid. Dyson has used the viscoelastic behaviour of liquids, measured under low amplitude oscillatory conditions, as a model for the variation of the viscosity in steady shear as a function of shear rate. Specifically, he assumes that the fall of viscosity with increasing shear rate follows the same form as the fall in the dynamic viscosity $\eta'(\omega)$ with increasing angular frequency ω. This approach, when combined with thermal effects, provides the basis for a fairly satisfactory explanation of the behaviour at high sliding speeds, but does not explain the variation of viscosity with rolling speed.

An explanation of this effect has been proposed by Fein (1967) and further developed by Harrison and Trachman (1972), which involves the time taken for a liquid to reach equilibrium following a rapid change in pressure. If this time is of the same order, or longer, than the residence time of the lubricant in the contact zone then the viscosity of the lubricant will be less than the equilibrium value associated with the pressure in the contact, and the effective viscosity will fall with increased rolling speed as the residence time is reduced further. By choosing a suitable value for the relaxation time associated with the compressional viscoelastic behaviour of the lubricant, Harrison and Trachman were able to obtain good agreement with the viscosity-rolling speed behaviour measured by Johnson and Cameron (1967). Paul and Cameron (1974) have recently observed directly the increase in viscosity with time following a sudden pressure increase.

An alternative explanation has been proposed by Adams and Hirst (1973). They suggest that an apparent reduction of the effective viscosity with rolling speed is due to changes in the pressure distribution over the contact area. These changes are caused by shear-rate induced viscosity changes in the inlet and outlet regions of the contact. A third alternative is proposed by Johnson and Roberts (1974). They have used a novel rolling-contact apparatus in which the axes of rotation of the rollers can be tilted or skewed relative to each other. It is then possible to distinguish between viscous and elastic behaviour in the lubricant. Their results indicate that above a critical pressure, which depends on the rolling speed, the liquid response becomes predominantly elastic, rather than viscous. They suggest that the apparent fall in viscosity with rolling speed observed in a conventional disc machine is a consequence of regarding the traction force between the rollers as arising from viscous forces only, and neglecting the contribution of the shear modulus of the liquid.

There are thus considerable differences in opinion regarding the causes of the observed behaviour in rolling contact lubrication. A common feature, however, is the treatment of the lubricant as a viscoelastic material, and a knowledge of liquid properties in both shear and compression over wide ranges of time scale, pressure and temperature will be essential to a better understanding of elasto-hydrodynamic lubrication.

Chapter 2 is devoted to a review of the equilibrium liquid properties of viscosity and density, and their variation with temperature and pressure. Many empirical equations have been used to describe these variations, but the range of temperatures and pressures over which they are valid is often very limited. The equations discussed here are those which have been found to apply over a wide range of conditions and which can be used to make predictions of the liquid properties outside the normal range of measurement.

An introduction to the theory of linear viscoelasticity is given in Chapter 3. The propagation of shear waves in liquids is analysed and the relationships between the complex frequency-dependent parameters of modulus, compliance and impedance are derived. Simple mechanical models, the Maxwell and Voigt elements, are used to illustrate the response of materials to the application of both transient and oscillatory stress and deformation. This approach is more easily understood than a rigorous mathematical treatment, and leads directly to the general concept of a continuous distribution of processes defined in terms of a spectrum of relaxation or retardation times. This chapter concludes with a discussion of the method of reduced variables, by which experimental data obtained over a range of temperatures and pressures may be referred to a single reference state, enabling the complete range of viscoelastic behaviour to be explored with apparatus of limited frequency range.

Chapter 4 is concerned with the various high frequency techniques which may be used to determine the dynamic mechanical properties of liquids. Techniques involving the generation of shear and longitudinal waves can be used within the frequency range 10^5 to 10^9 Hz, while optical diffraction techniques enable measurements to be obtained in the range 10^9 to 10^{10} Hz. Longitudinal waves generally propagate in a liquid with much less attenuation than shear waves. Hence, longitudinal wave techniques use propagation methods, in which the absorption coefficient and velocity of the wave in the liquid are measured. In shear wave techniques the wave is usually propagated in a low-loss solid material, either as a free-space wave or as a guided wave, and interaction with the liquid occurs at the surface of the solid. The mechanical impedance of the liquid is then determined from the behaviour of the wave at the solid-liquid interface.

The results obtained from shear wave measurements on super-cooled liquids and liquid mixtures are reviewed in Chapter 5. It is convenient to describe the experimental results in terms of empirical equations, such as the B.E.L. relaxation equation, and the Davidson–Cole retardation equation. The relative merits of these equations are discussed with reference to the accuracy with which the parameters can be determined, and the extent to which they can describe the complete viscoelastic behaviour of a liquid.

The response of liquids to longitudinal waves is treated in Chapter 6, together with a brief survey of dielectric relaxation. This chapter concludes with a comparison of dielectric, shear and structural behaviour, and with a discussion of the relations between these observed properties and the fundamental molecular motions which occur in the liquid state.

Throughout this book it is assumed that the elastic waves are of low amplitude, so that the response of the liquid to a wave is linear and independent of the amplitude of the wave. The total energy flow from the wave to the liquid is also low and no permanent changes are produced, the equilibrium state in the liquid being only slightly perturbed. Similarly only linear behaviour in steady flow and in transient experiments is considered, non-linear effects such as the appearance of normal forces, or shear-thinning, being outside the scope of this book. The mathematical content has been kept to the minimum necessary for the analysis and understanding of the experimental data, and no attempt has been made to include largely mathematical topics such as the use of constitutive equations to describe liquid rheology, or the general theories of the liquid state. References to the original literature have been included where appropriate and should provide an adequate guide to a more detailed study of the subject.

2. The Liquid State

A liquid is usually defined by the following concise and simple definition – 'A liquid is a fluid which if placed in a closed vessel at once conforms to the shape of the vessel without necessarily filling the whole of its volume' (Rowlinson, 1969).

The first property, that of being unable to maintain its shape, distinguishes a liquid from a solid. The second property, that of forming a free surface and not completely filling a container, distinguishes a liquid from a gas. These two properties are adequate for the identification of the commonly experienced liquids such as water, alcohol and many other simple organic liquids. For such materials the liquid state is clearly intermediate between the solid and gaseous states. At sufficiently low temperatures and high pressures the material exists as a solid; at high temperatures and low pressures as a gas (in the absence of chemical decomposition). The transitions between the states are marked by the phenomena of freezing and boiling, occurring at well defined temperatures.

For materials such as pitch, glass and many polymers, this simple definition is inadequate, however. At room temperature such materials often appear to be solids, apparently maintaining their shape indefinitely. On raising the temperature melting occurs and the material flows like a liquid. The transition from solid to liquid does not now occur at a well defined melting temperature, though, and is not accompanied by the absorption of a latent heat and the discontinuity in density normally associated with the phenomenon of melting. Also, at temperatures sufficiently high for flow to be possible, the flow is often sluggish, the material having such a high viscosity that it is not possible for it to "at once" conform to the shape of the containing vessel, as required by the simple definition.

Such a material, characterised by high values of viscosity and no definite melting point, is described as a "supercooled" liquid, or, in the solid state when the viscosity is so high that the rate of flow is imperceptible even over long times, as a glass. A glass has the external appearance of a solid, but maintains the molecular structure of a liquid, lacking the long range order and regular structure of a crystalline solid. The usual lower temperature limit of the liquid state, the melting point, at which crystallisation occurs in simple liquids, does not apply in the case of supercooled liquids, as the material stays in the liquid state as the temperature is lowered. A characteristic of supercooled liquids is a high viscosity in the region of the freezing point. For example, glycerol is found to have a freezing point of $17.9°C$, below which crystallisation occurs at a slow rate. At this temperature the viscosity is about 1.7 Pa s, or about 1000 times the viscosity of water at its freezing point. If glycerol is cooled quickly below its freezing point the viscosity increases substantially and the molecular rearrangements necessary for the formation of a crystalline lattice are slowed down so that the rate of crystallisation becomes extremely slow, and the material stays in the liquid state. Eventually, at sufficiently low temperatures, the motions are virtually at a standstill, and the random configuration of the liquid is "frozen in" permanently in the glass (see Pryde, 1966, Ch. 3).

The upper temperature limit of the liquid state is the critical temperature, for both simple and supercooling liquids. The critical temperature is that temperature at which the liquid and gas phases can no longer be distinguished apart, regardless of the applied pressure. At atmospheric pressure the upper temperature limit is, of course, the boiling point.

2.1 Density

The values of the density (mass per unit volume) of the more common liquids are close to the values for crystalline solids, the change in the density when a solid melts being on average only about 10%. This closeness in the values for solids and liquids is indicative of the similarity in the molecular packing in these two states. In the S.I. system of units, density is expressed in $kg\ m^{-3}$, the value for water at $3.98°C$ being taken as $1000.0\ kg\ m^{-3}$. The densities of most

2. THE LIQUID STATE

organic liquids lie close to this value, in a range from about 600 kg m^{-3} to about 3000 kg m^{-3}. Most mineral, vegetable and animal oils have densities between 850 kg m^{-3} and 950 kg m^{-3}. Other commonly used density units are:

in c.g.s. units, g cm^{-3}. (1 g cm^{-3} = 10^3 kg m^{-3})
in British units, lb ft^{-3}. (1 lb ft^{-3} = 16.018 kg m^{-3})

The extensive data tabulated by the A.P.I. Research Project 44, (1953) and Timmermans (1965) shows that for many liquids the density varies linearly with temperature, following an equation of the form

$$\rho = \rho_r(1 - a(t - t_r)) \qquad (2.1)$$

where ρ_r is the density at the arbitrary reference temperature t_r. The coefficient a has a value of around 10^{-3} for reference temperatures around 0°C. The deviation from linearity is usually found to be less than 1 part in 10^3 over a range of temperature of some 200K. The results of Greenwood and Martin (1952), and Barlow et al. (1966) show that this linearity extends well into the supercooled region.

The variation of density with pressure may be expressed in several ways, of which the compressibility and the bulk modulus are the most common. The data are usually presented in terms of the volume relative to that at atmospheric pressure, V/V_0, as a function of the applied pressure. The ratio of the change in volume to the reference volume, $(V_0 - V)/V_0$, is termed the compression. The compressibility is defined as the fractional change in volume with pressure, $-(dV/dP)/V$, at a specified pressure, determined from the slope of the volume–pressure curve. The secant bulk modulus refers to the overall change in volume between two pressures, $V_0(P - P_0)/(V - V_0)$, whereas the tangent bulk modulus, the inverse of the compressibility, is given by $-V(dP/dV)$, and is a measure of the incremental resistance of the liquid to being compressed.

Data on the compressibility of many liquids are to be found in the publications of P. W. Bridgman in the Proceedings of the American Academy of Arts and Sciences, or in abbreviated form in "The Physics of High Pressure" (Bridgman 1949).

Figure 2.1 shows the behaviour typical of most liquids, the volume decreasing steadily, but at a decreasing rate as the pressure is increased. For a large number of organic liquids at room tempera-

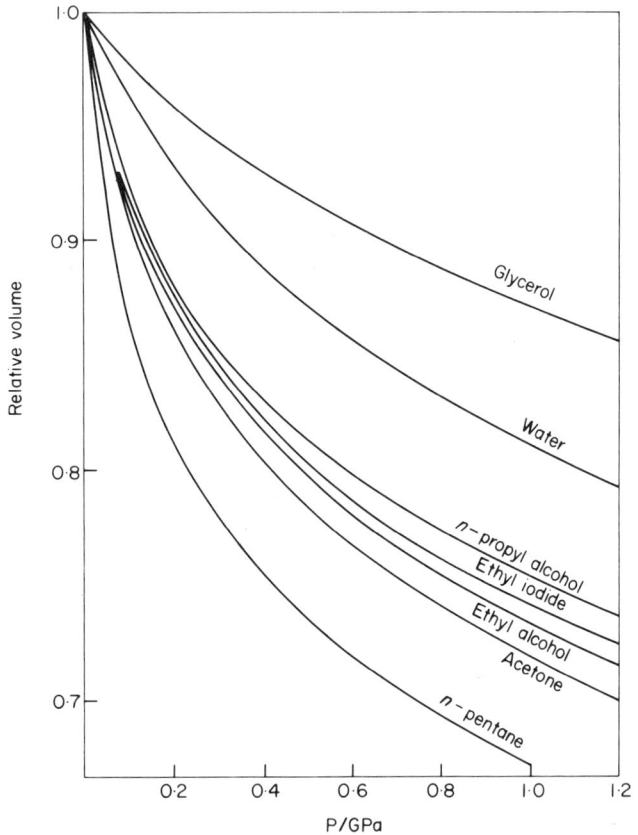

FIG. 2.1. Relative volume of several liquids plotted as a function of pressure at a temperature of 50°C (data from Bridgman (1949)).

ture, the volume is reduced by between 20% and 30% by the application of a pressure of 1 GPa. Glycerine is rather less compressible, losing only 13% of its original volume. In contrast to these liquids mercury loses less than 4%. The compressibility of all liquids decreases considerably with increasing pressure, the value at 1 GPa being reduced to about 1/15 of the atmospheric value in many cases. Although large differences in compressibility between different liquids occur at atmospheric pressure (up to a factor of 10), at a pressure of 1.2 GPa compressibilities are found to lie within 20% of $0.08\,(\text{GPa})^{-1}$, with very few exceptions. Further data on the compressibility of liquids are available in the Research Reports 42 and 44 of the American Petroleum Institute (1953, 1967) (pure hydrocarbons) and in the A.S.M.E. "Pressure–Viscosity Report" (1953) (mineral oils and some synthetic lubricants).

Many units of pressure are in use, but much high pressure data are presented using $kg_f \, cm^{-2}$, atm, or psi. The relations between the more common units are given in Table 2.1.

Hayward (1967) has made a comparative study of the various equations that have been used to express the compressive properties of liquids. Many equations reduce to the linear secant bulk modulus equation,

$$\bar{K} = \frac{V_0(P - P_0)}{V_0 - V} = K_0 + m_2(P - P_0).$$

This equation, giving a linear variation of the secant bulk modulus $\bar{K}$ with pressure, is easily fitted to experimental data, and adequately describes the behaviour of many liquids up to pressures of the order of 100 MPa. Over larger pressure ranges, higher order terms can be added, and a cubic equation of the form $\bar{K} = K_0 + m(P - P_0) - n(P - P_0)^2 + q(P - P_0)^3$ will describe the behaviour of organic liquids up to pressures of 1 GPa.

The linear secant bulk modulus equation may be written in terms of density ρ as

$$\frac{1}{\rho} = A + \frac{B}{K + (P - P_0)}.$$

2.2 Viscosity

The physical property which characterizes the flow resistance of a simple fluid is the viscosity. The law relating viscosity to the shear stress causing the flow was postulated by Isaac Newton in 1687: (Reiner, 1960) "Hypothesis: that the resistance which arises from the lack of slipperiness of the parts of a liquid, other things being equal, is proportional to the velocity with which the parts of the liquid are separated from each other". Newton's "lack of slipperiness" is what is now called viscosity. If the liquid is contained

TABLE 2.1 Units of pressure

Unit below equals unit across multiplied by number in table	dyne cm^{-2}	Bar	atm	kg$_f$ cm^{-2}	Pascal (N m^{-2}) Pa	lb$_f$ in^{-2}
dyne cm^{-2}	1	10^{-6}	0.987 × 10^{-6}	1.020 × 10^{-6}	0.1	1.450 × 10^{-5}
Bar	10^6	1	0.987	1.020	10^5	14.50
Atmosphere (atm)	1.013 × 10^6	1.013	1	1.033	1.013 × 10^5	14.70
kg$_f$ cm^{-2}	0.981 × 10^6	0.981	0.968	1	0.981 × 10^5	14.22
Pascal (Pa = N m^{-2})	10	10^{-5}	0.987 × 10^{-5}	1.020 × 10^{-5}	1	1.450 × 10^{-4}
lb$_f$ in^{-2} (psi)	6.895 × 10^4	6.895 × 10^{-2}	6.804 × 10^{-2}	7.031 × 10^{-2}	6.895 × 10^3	1

2. THE LIQUID STATE

between two parallel planes of area A, moving under the action of a force F with a relative velocity V and separated by a distance h, the value of the force is proportional to the velocity gradient in the liquid. The shear stress σ is given by $\sigma = F/A = \eta V/h$, where the constant of proportionality η is the viscosity. (See Fig. 2.2). The velocity gradient may be equated to the rate of shear $\dot\gamma$. If the angle of shear of the liquid is defined as $\gamma = dx/dy$, the shear rate is $\dot\gamma = dv_x/dy$ where $v_x = dx/dt$. Then Newton's Law may be written

$$\sigma = \eta\dot\gamma \qquad (2.2)$$

The quantity viscosity divided by density is termed the kinematic viscosity, the usual symbol is ν. The kinematic viscosity is the quantity which is measured directly in some types of viscometer.

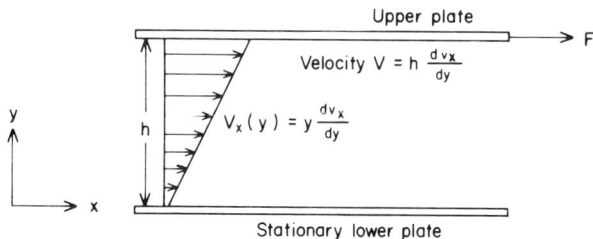

FIG. 2.2. Velocity gradient in plane laminar flow.

2.2.1 VISCOSITY UNITS

In the S.I. system shear stress is measured in units of Pa($= $ N m^{-2}), and velocity gradient in units of m s^{-1}/m $=$ s^{-1}. The unit of viscosity is thus Pa s or m^{-1} kg s^{-1}. The kinematic viscosity $\nu = \eta/\rho$ is then measured in m^2 s^{-1}.

Most viscosity data in the literature are expressed in the c.g.s. unit dyne cm^{-2}s, which is called the poise (P). The c.g.s. unit for kinematic viscosity, cm^2 s^{-1} is called the stoke (St). The submultiples centipoise (cP $=$ 0.01P) and centistoke (cSt $=$ 0.01 St) are also commonly used. In terms of S.I. units these quantities are given by

$$1P = 100 \text{ cP} = 0.1 \text{ Pa s}$$
$$1St = 100 \text{ cSt} = 10^{-4} \text{ m}^2 \text{ s}^{-1}$$

In British units viscosity is expressed in several different ways. If the shear stress is expressed in poundals ft^{-2} then viscosity has the unit poundals ft^{-2} s or lb$_m$ ft^{-1} s^{-1} (= 1.488 Pa s). When stress is measured in lb$_f$ ft^{-2}, equation (2.2) may be written

$$\sigma g_c = \eta \frac{dv_x}{dy}$$

where g_c, the "gravitational conversion factor", has the value 32.174 poundals/lb$_f$, or (lb$_m$/lb$_f$)(ft sec^{-2}). Alternatively, η/g_c may be taken as a measure of viscosity and has the units lb$_f$ ft^{-2} s (= 47.88 Pa s). The related unit lb$_f$ in^{-2} s is called the Reyn (= 6.89 x 10^3 Pa s); the stress is then measured in lb$_f$ in^{-2} (psi).

Several industrial viscometers characterise the viscosity of a liquid in terms of the time taken for a specific quantity of liquid to flow out of a container through an orifice of a given size. Such instruments, commonly used to characterise lubricants and paints, give rise to a series of "viscosity units" which are not related in any simple way to either the viscosity or the kinetic viscosity; tables and charts giving approximate relations between the kinematic viscosity and the efflux times of the different instruments are available (Handbook of Chemistry, 1961; Stearns, 1970).

The most common units, and the approximate range of the instruments are given in Table 2.2.

TABLE 2.2 Approximate ranges of efflux viscometers (Stearns, 1970)

Instrument		Unit	$10^6 v/m^2$ s^{-1}
Saybolt Universal	(American)	seconds	1–400
Saybolt Furol		seconds	400–4000
Redwood No 1	(British)	seconds	1–500
Redwood No 2		seconds	500–5000
Engler	(German)	degrees	1–1500

2.2.2 VISCOSITY VALUES

In Table 2.3 some experimental viscosity data are given for some materials which are liquid at atmospheric pressure. Most simple organic liquids have viscosities in the range 10^{-4} to 10^{-3} Pa s at room temperature. Higher values of viscosity are found in liquids which supercool; this may occur at low temperatures for low molecular weight liquids (e.g. glycerol) or at high temperatures for glasses and many polymers.

TABLE 2.3   Approximate viscosities of some liquids at atmospheric pressure

Material	Temperature/°C	Viscosity/Pa s
Ethyl Ether[a]	20	2.43×10^{-4}
Benzene[a]	20	6.5×10^{-4}
Water[a]	20	1.0019×10^{-3}
Ethyl Alcohol[a]	20	1.19×10^{-3}
Mercury[a]	20	1.55×10^{-3}
Crankcase Oil, SAE 10W[b]	38	3.5×10^{-2}
Crankcase Oil, SAE 30[b]	38	0.10
Crankcase Oil, SAE 50[b]	38	0.24
Gear Oil, SAE 90[b]	38	0.27
Castor Oil[a]	20	0.96
Glycerol[a]	20	1.50
Glycerol[a]	−20	134
Polydimethysiloxane, M.Wt. = 6×10^3[c]	20	0.10
Polydimethylsiloxane, M.Wt. = 6×10^4[c]	20	100
Polystyrene, M.Wt. = 1.17×10^5[d]	183	2.9×10^3
Polystyrene, M.Wt. = 2.42×10^5[d]	183	3.3×10^4

Data taken from: [a] American Institute of Physics Handbook (1963) [b] Lee and Booser (1970) [c] Barlow et al. (1964) [d] Stratton (1966).

2.2.3   NON-NEWTONIAN BEHAVIOUR

Liquids in which the viscosity is a constant independent of the shear rate are called Newtonian liquids. Liquids of low molecular weight, and low viscosity, are typically Newtonian up to shear rates of at least 10^6 s^{-1}. However, many materials of high molecular weight, especially molten polymers and polymer solutions, suspensions, pastes, and other complex systems, show deviations from Newtonian behaviour, i.e., the coefficient of viscosity η in equation (2.2) is not independent of shear rate. With increasing shear rate, most polymer melts and solutions show a decrease in the "apparent viscosity", defined as the ratio of shear stress to shear rate at a particular shear rate. Such behaviour is described as "shear thinning" if the fall in the value of the viscosity is dependent only on the magnitude of the shear rate, and the value returns to the original value immediately the shearing is stopped. If the viscosity remains at a lower value after the shearing has stopped, and only returns to the original value after the material has stood for some time, the material is said to be "thixotropic": "non-drip" or "jelly" paint is a common example of a thixotropic material.

Materials in which the apparent viscosity increases with shear rate are termed "shear-thickening" or "dilatant". This phenomenon is less common than shear thinning, but is observed in some suspensions of solid particles. Materials which have a paste-like consistency may show a yield value — a shear stress below which flow does not occur. Materials which thereafter have a linear flow curve (stress vs. shear rate) are called "Bingham materials".

2.2.4 VARIATION OF VISCOSITY WITH TEMPERATURE

The viscosity of liquids is found to decrease with increasing temperature. Near the boiling point, where the viscosity of a simple organic liquid is typically 10^{-4} Pa s, the viscosity is found to vary as an exponential function of the absolute temperature according to the equation

$$\eta = A e^{B/T} \qquad (2.3)$$

or $\qquad \ln(\eta/A) = B/T$

where A and B are constants. This empirical equation is usually referred to as the Andrade or Arrhenius equation (Andrade, 1930). This equation often fails for liquids near their melting points and is not applicable to supercooled liquids. Many other empirical equations have been used, often for specific types of liquid and over restricted temperature ranges. Two equations widely used for lubricating oils are the Walther (1931) and Roelands (1966) equations.

The Walther equation is

$$\log(10^6 \, \nu/\mathrm{m}^2 \, \mathrm{s}^{-1} + 0.6) = C(T/K)^{-m} \qquad (2.4)$$

where m and C are constants. This equation forms the basis of the ASTM viscosity–temperature chart, No D 341 (ASTM, 1970), used for estimating the viscosity of oils. The Roelands equation, which is found to hold for a wide range of different types of lubricant, is

$$\log(\eta/\text{N s m}^{-2}) = C(T/\text{K} - 138)^{-m} - 4.2 \qquad (2.5)$$

where m and C are constants. Both of these equations find their main application in lubricants of fairly low viscosity (below 10 Pa s and at temperatures above $-40°\text{C}$).

Another empirical equation which has been widely used is

$$\ln(\eta/A) = B(T/\text{K})^{-m} \qquad (2.6)$$

where A and B are constants and m is a constant chosen arbitrarily to give the best straight line fit to the data. A value of m of 4 has been used to describe the viscosity of lubricating oils and supercooled liquids by Barlow and Lamb (1959) and Barlow et al. (1966). Litovitz (1952) has used this equation with $m = 3$ to describe the viscosity of several associated liquids over a temperature range of some 100K above and below their melting points. All of the above two parameter equations are convenient for the immediate analysis and interpolation of experimental data, as the results can be readily plotted, but a straight line is obtained over a limited temperature range only, and extrapolation outside the experimental range cannot be carried out with any degree of confidence.

An alternative equation for the temperature variation of viscosity is based on the concept of free volume. Doolittle (1951) first proposed an empirical equation using this concept, to represent viscosity measurements over a wide temperature range, expressed in the form

$$\ln\left(\frac{\eta}{A}\right) = B\frac{v_0}{v_f} \qquad (2.7)$$

where the specific volume is $v(= v_0 + v_f)$, v_f is the free volume and v_0 is the volume associated with the liquid when in its state of closest molecular packing. Thus v_f represents the volume available in the liquid for the translation of molecules. When v_f is zero the viscosity will be infinite and $v = v_0$. If this is assumed to occur at some (low) temperature T_0, the density equation (2.1) may be rewritten using T_0 as the reference temperature to give

$$\rho = 1/v = \rho_0(1 - a_0(T - T_0))$$

where $\rho_0 = 1/v_0$. This may be rearranged to give

$$\frac{v_0}{v_f} = \frac{1}{a_0(T-T_0)} - 1$$

Then equation (2.7) may be written

$$\ln\left(\frac{\eta}{A'}\right) = B'/(T-T_0) \qquad (2.8)$$

This equation, which will be referred to as the modified free volume equation, has also been used empirically by Fulcher (1925) and Tamman and Hesse (1926).

Figure 2.3 shows a schematic representation of the variation of specific volume with temperature according to these ideas of free volume. The temperature T_0 defines a thermodynamic second-order transition, the fundamental reference temperature for the liquid state. Below T_0 the material would exist in a state of closest random packing of the molecules as an equilibrium glass. If the volume is measured experimentally as a function of temperature, however, a change in the slope of the graph is found to occur at a temperature corresponding to a viscosity of about 10^{12} Pa s, not at T_0. This temperature is the glass transition temperature, T_g. Above T_g the material is a liquid, below it behaves as a glass. The value of the glass transition temperature depends on the time scale of the experiment. If a sufficient time is allowed for equilibrium to be reached when cooling the liquid, the transition occurs at a lower temperature. This effect is shown, for example, by the work of Kovacs (1958) on the dependence of T_g in polymers on the time scale of the experiment (Fig. 2.4).

The glass transition temperature then represents a practical limit to equilibrium measurements, the time required to attain equilibrium becoming significantly longer than the time scale of the experiment at lower temperatures. The material exists at temperatures below T_g in a non-equilibrium state, the structure of the liquid at T_g being "frozen-in". In long chain polymers, a series of glass transitions have been observed as the motions of side-branches and other smaller parts of the molecule are successively immobilized at temperatures below the main T_g. An average relaxation time τ for a liquid is given

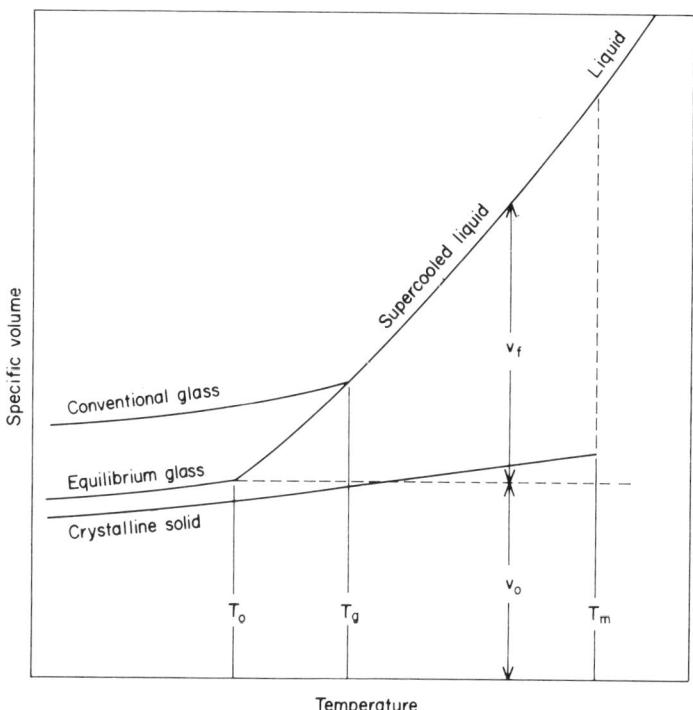

FIG. 2.3. Schematic representation of the variation of specific volume with temperature for a crystalline solid, an equilibrium glass, a conventional (non-equilibrium) glass and a liquid. T_m is the melting temperature of the crystalline solid, T_g is the conventional glass transition temperature, T_0 is the true second order transition temperature and v_0 is the temperature-independent occupied volume. (From Plazek and Magill (1966)).

by the ratio of the viscosity to the limiting rigidity modulus at high frequencies, G_∞, $\tau = \eta/G_\infty$ (see § 3.2). The value of G_∞ is typically 10^9 Pa, and at T_g the viscosity is of the order of 10^{12} Pa s. The relaxation time of the liquid is thus of the order of 10^3 s, or about 15 minutes, which is comparable with the duration of a normal experiment. The use of techniques involving short time scales enables glass transition effects to be studied at temperatures well above the conventional T_g; a description of such techniques and the results obtained are the subject matter of later chapters.

The modified free volume equation (2.8) is used in practice by

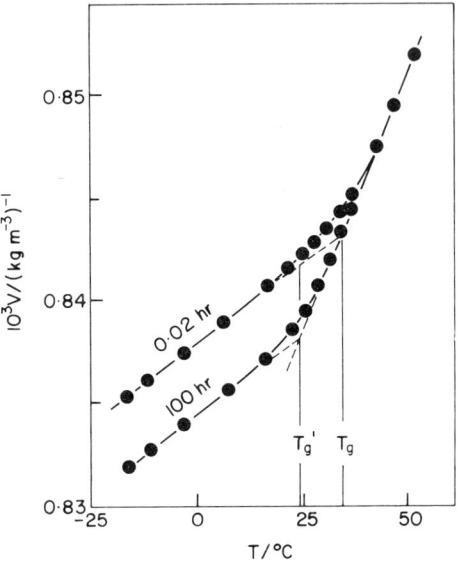

FIG. 2.4. Specific volume of amorphous polyvinyl acetate as a function of temperature. Volumes were observed at 0.02 hr and 100 hr after cooling rapidly from a high temperature. T_g and T_g' are the glass transition temperatures corresponding to these two different experimental time scales (after Kovacs, 1958).

obtaining values of A', B' and T_0 which give a best fit to the experimental data. This is conveniently carried out using a digital computer to minimise the sum of the squares of the differences between the experimental and the predicted viscosity values. The equation can be used, after fitting to experimental data covering viscosities up to about 100 Pa s, to predict the glass transition temperature by inserting the value 10^{12} Pa s for the viscosity. For many supercooled liquids the values so obtained are found to be within ±5K of the values determined by differential thermal analysis (Barlow *et al* 1969b). This critical test demonstrates the utility of the free volume equation in describing the viscosity—temperature variation of supercooled liquids in the high viscosity region. The upper temperature limit of the equation, T_A, has been found to be at a free volume of between 10% and 16%: thereafter the Arrhenius equation

(2.3) gives a more satisfactory fit to the data. Macedo and Litovitz (1965) have suggested that the viscosity of a liquid can be described over the whole temperature range by combining equations (2.3) and (2.8) to give a hybrid equation of the form

$$\ln(\eta/A) = B/(T - T_0) + C/T \qquad (2.9)$$

However, separation of the data into two distinct regions, the high temperature region being described by the Arrhenius equation and the low temperature region being described by the free volume equation, is found to give a considerably better fit to the experimental results than the use of the hybrid equation.

It should be noted that the "W.L.F." equation (Williams et al 1955), widely used to describe the temperature variation of polymers, is identical in form to the modified free volume equation. In its usual form the W.L.F. equation is written

$$\log(\eta/\eta_g) = \frac{-C_1(T - T_g)}{C_2 + (T - T_g)}$$

where η_g is the viscosity at T_g and C_1 and C_2 are constants. This is identical with equation 2.8 if $C_1 C_2 = B'/2.303$, $C_2 = T_g - T_0$.

The modified free-volume equation (2.8) is found to be an adequate representation of the viscosity—temperature behaviour for many liquids in the supercooled region. In some liquids, however, a "discontinuity" in the viscosity behaviour is observed in the region of a temperature T_K which is at, or near, the melting point (Barlow et al. 1966). Below this temperature the observed viscosity values can be fitted to equation (2.8) with one set of values of the parameters A', B' and T_0. A similar fit can be obtained in the temperature range between T_K and T_A, the Arrhenius temperature, but a different set of values for the parameters is required. Above T_A the behaviour follows (by definition) the exponential variation given by equation (2.3). If equation (2.6) is used to represent the data, the graph consists of two straight lines intersecting at T_K: the sudden change in slope cannot be eliminated by variation of the parameter m.

This type of behaviour has been discussed in detail by Davies and Matheson (1966, 1967) in terms of the rotational freedom available to molecules. At temperatures above the Arrhenius temperature T_A, the probability that molecules can make a translational jump, from one site in the liquid to another, is determined solely by the availability of the necessary energy. This is the mechanism proposed in the rate theory of Eyring (1936) which predicts a viscosity proportional to exp (E/RT) where E is an activation energy of viscous flow (i.e. equation 2.3). Under these conditions molecules are free to rotate many times about at least two axes during the time between jumps. Liquids composed of spherical molecules, such as the liquid inert gases, liquid metals and molten salts which contain only monatomic ions, show Arrhenius behaviour throughout the normal liquid range.

At temperatures below T_A the viscosity behaviour becomes controlled by free-volume considerations rather than by just the available energy. The free-volume equation has been derived theoretically by both Cohen and Turnbull (1959) and by Kumar (1963). It is assumed that molecular translation takes place by the diffusion of molecules into defects, or holes, in the liquid structure. The continuous random redistribution of the free volume occasionally results in the accumulation of sufficient free volume in one place — a hole — to allow a molecule to diffuse into it from an adjacent site. Davies and Matheson argue that a molecule whose rotation is restricted, by virtue of its shape, will diffuse less readily than one which can rotate freely and can thus re-align itself in order to move more easily past its neighbours. They therefore postulate that "the deviation of the temperature dependence of viscosity from Arrhenius behaviour occurs when molecular rotation about two axes becomes restricted on the translational time scale".

If the shape of the molecule is such that no free rotation is possible below T_A then one region of non-Arrhenius behaviour will be observed, and equation (2.8) can be used to describe the whole supercooled region. If free rotation about one axis is possible between T_A and T_K, with this motion being restricted at temperatures below T_K, then two regions of non-Arrhenius behaviour will be observed. Molecules such as di-methyl phthalate, di-iso-butyl phthalate and several simpler benzene derivatives are supercooling liquids which have only short side-groups attached to a benzene ring. Rotation about one axis is possible between T_A and T_K resulting in two distinct regions of non-Arrhenius behaviour. Molecules with larger side-groups, such as di-n-butyl phthalate and di(2-ethyl hexyl) phthalate, can not rotate freely below T_A and equation (2.8) describes the whole supercooled region with one set of parameters.

At the glass transition temperature, when $\eta = 10^{12}$ Pa s, any rotational motion will have been restricted by the lack of free volume, and so the observed T_g corresponds to the value predicted by equation (2.8) for the lower temperature region of the liquid, below T_K. If viscosity data are being extrapolated to lower temperatures it is therefore necessary to ensure that only data obtained below T_K are used when determining the parameters of equation (2.8). (Carpenter *et al.* 1967).

Laughlin and Uhlmann (1972) have studied the behaviour in the high-viscosity region of some simple organic glass-forming liquids (salol, α-phenyl-*o*-cresol, *o*-terphenyl and tri-α-naphthylbenzene). Viscosities in the range 10^7 to $10^{12.5}$ Pa s were measured using a beam bending technique. The liquids were cast into cylindrical rods in low temperature graphite moulds, and then the deflection of the rod was measured when it was subjected to a bending load. A falling sphere viscometer was used for viscosities below 10^7 Pa s. The free-volume equation was found to provide a good description of the data for viscosities below 10^3 Pa s. At low temperatures, for viscosities higher than this, the data could be best described by the Arrhenius equation (2.8), although slight curvature occurred in the log (η) vs $1/T$ plot for some liquids. The activation energy for viscous flow was in all cases considerably higher than the values associated with the high temperature Arrhenius region. Laughlin and Uhlmann suggest that while the flow mechanism near the glass transition temperature can be described in terms of transport over a potential energy barrier, an extensive degree of co-operation motion must be involved. Magill and Li (1973) have found that the empirical equation

$$\log(\eta/\eta_s) = A \left[\exp(BT_g/T) - \exp(B/(1 + \phi))\right]$$

describes the viscosity variation of a range of organic glass-forming liquids and polymers over the complete range of viscosities (over 14 orders of magnitude). In this equation η_s is the viscosity at a reference temperature T_s given by $T_s = T_g(1 + \phi)$. A value of 0.24 for ϕ is found to give a good fit for all materials, and values of $A = 0.0111$, $B = 7.19$ enable the data for seven materials to be described by this equation to within a factor of two in the viscosity values.

If the data are to be fitted to a higher order of accuracy than this, no single simple equation can provide an entirely satisfactory description of flow over the full range of viscosities. Even when an equation is found to describe an extensive portion of the viscosity range, extrapolation outside the range of applicability can lead to considerable error.

2.2.5 VARIATION OF VISCOSITY WITH PRESSURE

The viscosity of most liquids increases markedly with the application of pressure. Water is anomalous, in that at temperatures below 20°C the viscosity first decreases, before increasing, with increasing pressure (Bridgman 1949). For other liquids, the viscosity is an exponential function of pressure, for pressures up to about 100 MPa. Equation (2.10) describing this behaviour is usually called the Barus equation after C. Barus who proposed it as an empirical approximation (Barus, 1893).

$$\eta = \eta_0 \exp(\alpha P) \qquad (2.10)$$

The "pressure–viscosity" exponent α is typically in the range 7 to 70 $(GPa)^{-1}$, and is temperature dependent, decreasing with increasing temperature. Within certain classes of liquids a correlation exists between α and the atmospheric pressure viscosity η_0. For example, in a wide range of mineral oils, Worster (1951) finds that

$$\alpha = [35.8 + 9.9 \log(\eta_0/Pa\ s)]\ (GPa)^{-1}$$

when η_0 is expressed in Pa s.

Over a larger pressure range, this simple exponential law is not adequate, as a plot of log viscosity versus pressure is curved, the viscosity increasing less rapidly at the higher pressures. In some liquids a point of inflection is observed, the value of the slope increasing at the highest pressures. This type of behaviour is shown in Fig. 2.5 for a silicone lubricating oil. It may be argued that in general this upturn in the curve is to be expected at the higher pressures, as at some sufficiently high pressure the viscosity will become infinite, in the absence of phase changes, as the available free volume is reduced to zero.

Several empirical equations describe the behaviour up to the point of inflection. Among these, that proposed by Roelands (1966)

2. THE LIQUID STATE

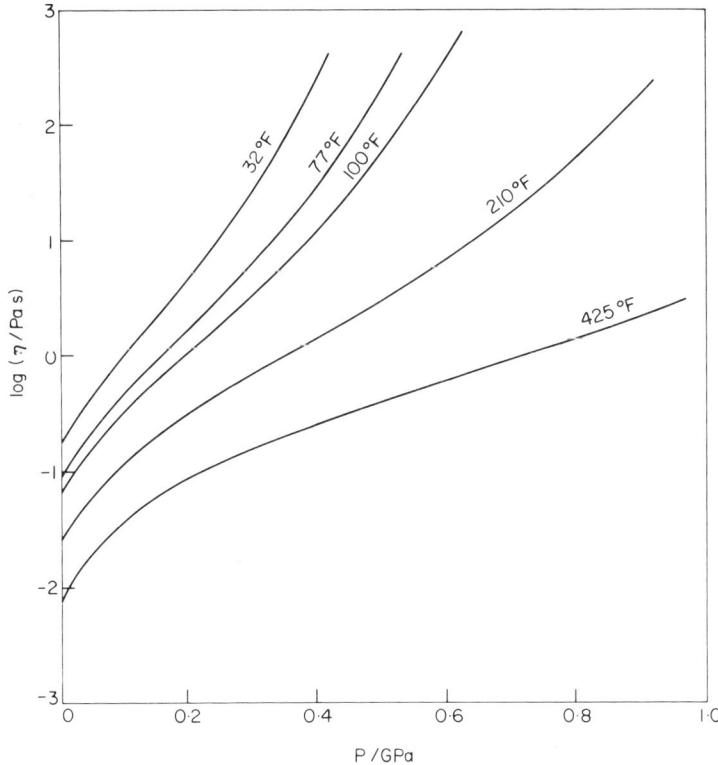

FIG. 2.5. Viscosity of a silicone oil plotted as a function of pressure at various temperatures (Silicone oil No. 9981LTNV-70, General Electric Co., from A.S.M.E. Pressure-Viscosity Report, 1953).

deserves mention in that the pressure–viscosity relationship is determined by the single parameter z:

$$\log(\eta/\text{Pa s}) + 4.2 = [\log(\eta_0/\text{Pa s}) + 4.2]\,[1 + 5.1P/\text{GPa}]^z \quad (2.11)$$

In equation (2.11), η_0 is the viscosity at atmospheric pressure. The parameter z is found to be independent of temperature, with a value in the range 0.3 to 1.5. With this equation experimental data on a wide range of liquids, including mineral oils, synthetic lubricants, pure hydrocarbons and a range of alcohols can be adequately represented up to pressures of the order of 500 MPa. Chu and Cameron (1962) have used a similar equation to represent the behaviour of paraffinic lubricating oils up to pressures of 1 GPa. For oils of viscosity greater than 6×10^{-3} Pa s their equation is

$$\log(10^3 \eta/\text{Pa s}) = \log(10^3 \eta_0/\text{Pa s}) \{11.0[\log(10^3 \eta_0/\text{Pa s})]^{-1/2} P/\text{GPa} + 1\}^{2/3} \quad (2.12)$$

where η_0 is the viscosity at atmospheric pressure.

To adequately describe the behaviour beyond the inflection point in the $\log \eta$ vs. pressure curve a more complicated expression must be used. For example, a double exponential equation of the form

$$\log(\eta/A) = \exp(BP) - C \exp(DP) \quad (2.13)$$

where A, B, C and D are constants at a given temperature, has been used by Irving and Barlow (1970). This equation describes the variation of viscosity over four orders of magnitude to an accuracy of better than 5%.

Equations which attempt to combine both the effects of pressure and temperature into a single equation are likewise rather cumbersome, and may involve as many as six parameters. Roelands (1966) has combined equations (2.5) and (2.11) to give a four parameter equation which is found to be valid over a temperature range of some 200K and up to pressure of 300–500 MPa for a wide range of liquids. This equation is

$$\log(\eta/\text{Pa s}) + 4.2 = C(T/K - 138)^{-m} (1 + 5.1P/\text{GPA})^z \quad (2.14)$$

where $\qquad z = B - D \log(T/K - 138)$

and $\qquad B$, C, D and m are the parameters.

3. Linear Viscoelasticity

The phenomenological theory of linear viscoelasticity attempts to describe the mechanical behaviour of a material in terms of time dependent, or frequency dependent, functions which relate the stress in the material to the deformation. Classical elasticity theory applies to solids in which Hooke's Law is obeyed, that is, the stress is always directly proportional to the instantaneous strain, but independent of the history of the strain (e.g. the rate of strain). Classical hydrodynamic theory applies to liquids in which, by Newton's Law, the stress is directly proportional to the instantaneous rate of strain but is independent of the strain itself.

These laws refer only to equilibrium or steady state conditions and do not describe transient effects. Also, as with the more general viscoelastic response functions described in later sections, all inertial effects are neglected. Many materials follow closely the behaviour specified by these laws under conditions of infinitesimal strain or infinitesimal rate of strain, but under different conditions deviations from these classical patterns of behaviour often occur. When finite strains are applied to solids, the ratio of stress to strain may no longer be independent of the amount of strain. Similarly, with finite rates of strain, deviations from Newtonian behaviour may be observed, as discussed in Section (2.3.3). The point at which the behaviour can no longer be considered infinitesimal depends very much on the particular material and also on the level of accuracy required from the measurements. Such large-strain non-linear behaviour lies outside the scope of this book.

Even when both strain and strain rate are infinitesimal, it is often impossible to characterise a material by either of the two classical types of behaviour. Substances which exhibit both solid-like and liquid-like properties are said to show viscoelastic behaviour. Some authors (Reiner, 1958; Oldroyd, 1956; Blair, 1969) distinguish between viscoelastic materials (solids having some viscous, or dissipative, properties), and elastico-viscous materials (liquids having some elastic, or energy storing properties), but this distinction will not be maintained here. The term *viscoelastic* will be used to describe the properties of any material which, under the appropriate conditions, is able both to store energy in elastic deformation and also dissipate energy as heat. If the strain and the rate of strain are kept sufficiently small so that, in a given experiment, the ratio of stress to strain is a function of time (or frequency) only, and independent of the stress level, the material is said to show linear viscoelastic behaviour. Linear behaviour is easily obtained in dynamic (oscillatory) experiments where the amplitude of deformation is usually extremely small. Situations involving continuous flow or transient deformation require more care to ensure that the response is in the linear region.

A considerable body of literature exists concerning the mathematical aspects of the phenomenological theory of linear viscoelasticity (Alfrey and Doty, 1945; Leaderman, 1949; Nolle, 1950; Gross, 1953). It is felt more appropriate to develop the subject here in terms of simple mechanical models, however, with no attempt at a mathematically rigorous treatment. While this results in the overall structure and the elegance of the subject being somewhat obscured, the model approach is easier to understand and more readily related to the physical behaviour of the materials. For linear behaviour the results are equivalent to those derived using the more mathematical method.

3.1 Propagation of Plane Waves in a Liquid

In most of the high frequency techniques used for measuring the viscoelastic properties of liquids, a plane shear wave is propagated into the liquid. The motion will be assumed to be sinusoidal with the displacement in the x direction and the wave propagating in the y direction. All points on any plane wave front parallel to the x-z plane are assumed to have the same displacement at any instant in time. Then considering an infinitesimal rectangular parallelopiped of dimensions dx, dy and dz, with shear stresses acting on the faces normal to the y direction, as shown in Fig. 3.1, the net shear stress on the element is

$$(\sigma + \delta\sigma) - \sigma = \frac{\partial \sigma}{\partial y} dy$$

and the force is
$$f = \frac{\partial \sigma}{\partial y} dx\, dy\, dz$$

The equation of motion for the element is then

$$\rho\, dx\, dy\, dz\, \frac{\partial^2 \xi}{\partial t^2} = \frac{\partial \sigma}{\partial y} dx\, dy\, dz$$

or
$$\rho \frac{\partial^2 \xi}{\partial t^2} = \frac{\partial \sigma}{\partial y} \tag{3.1}$$

where ξ is the displacement in the x direction.

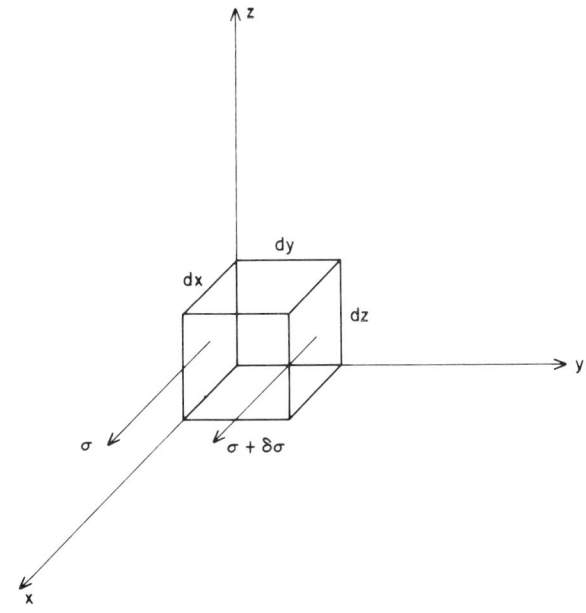

FIG. 3.1. Shear stresses on element of material.

The shear stress σ and shear strain γ are related by the shear modulus $G = \sigma/\gamma$, where the strain is given by $\gamma = \partial\xi/\partial y$. Then equation (3.1) becomes

$$\frac{\partial^2 \xi}{\partial t^2} = \frac{G \partial^2 \xi}{\rho \partial y^2} \qquad (3.2)$$

which is the wave equation for shear wave propagation in the y direction. Assuming the displacement varies sinusoidally with time, the solution for a wave propagating in the +ve y direction is

$$\xi = \xi_0 \exp[j\omega(t - y/c)]$$
$$= \xi_0 \exp[j(\omega t - \Gamma y)] \qquad (3.3)$$

where ω is the angular frequency, $c = \sqrt{(G/\rho)}$ is the velocity of propagation and $\Gamma = \omega/c$.

In a Hookean solid, G is a real quantity, the stress varies in phase with the strain, and the wave propagates with zero attenuation. In a Newtonian liquid, the stress is 90° out of phase with the strain, and the wave is attenuated as it propagates. The shear modulus, a pure imaginary quantity, may be determined from Newton's Law (Eq. 2.2). The strain rate $\dot{\gamma} = d(\gamma_0 \exp(jwt))/dt = jw\gamma$ so that the stress is given by

$$\sigma = \eta\dot{\gamma} = j\omega\eta\gamma$$

and
$$G = \frac{\sigma}{\gamma} = j\omega\eta.$$

Then
$$\Gamma = \omega\sqrt{\frac{\rho}{G}} = \sqrt{\frac{\omega\rho}{j\eta}} = (1-j)\sqrt{\frac{\omega\rho}{2\eta}}.$$

Thus equation (3.3) becomes

$$\xi = \xi_0 \exp\left[-\sqrt{\frac{\omega\rho}{2\eta}}\, y\right] \exp\left[j\omega\left(t - \sqrt{\frac{\omega\rho}{2\eta}}\, y\right)\right] \qquad (3.4)$$

and the wave travels with a phase velocity $c = \sqrt{(2\omega\eta/\rho)}$ and is attenuated with an attenuation coefficient $\alpha = \sqrt{(\omega\rho/2\eta)}$. The distance the wave travels before it is attenuated to 1/e of its original amplitude is

$$\lambda = \frac{1}{\alpha} = \sqrt{\frac{2\eta}{\omega\rho}} = \sqrt{\frac{\eta}{\pi f \rho}}$$

where f is the frequency in Hz.

In a viscoelastic material the stress and strain differ by a phase angle between these extreme values of 0° and 90°. The modulus is thus a complex quantity, denoted by $G^*(j\omega)$, and has real and imaginary components which are functions of frequency, i.e. $G^*(j\omega) = G'(\omega) + jG''(\omega)$. For a liquid with density $\rho = 10^3$ kg m^{-3} and viscosity $\eta = 0.1$ Pa s the value of γ for a frequency of 30 MHz is approximately 1 μm. Thus in this case the shear wave propagates over a very short distance only. Direct measurement of the velocity and attenuation of the wave, from which $G^*(j\omega)$ could be evaluated, is thus impossible, except at much lower frequencies. Instead, techniques are used which measure the effect of the liquid on the surface generating the wave. At the boundary between the generating surface and the liquid the values of both the shear stress and the displacement amplitude in the liquid must equal the values at the surface; measurements of the motion of the transducer generating the shear wave enable the liquid properties to be determined.

The quantity which is most easily measured is the mechanical shear impedance. This is defined by analogy with the specific acoustic impedance for a compressional wave, which is acoustic pressure divided by the particle velocity (Kinsler and Frey, 1962, p. 122). Thus a shear mechanical impedance of the liquid Z_L is defined as

$$Z_L = \frac{-\text{ Shear stress}}{\text{Particle velocity}}$$

$$= \frac{-\sigma}{\partial \xi / \partial t}$$

The negative sign is necessary because of the different sign of the displacements associated with pressure and stress; positive pressure results in compression, positive stresses produce expansion.

The particle velocity $\partial\xi/\partial t = j\omega\xi$. The shear stress is given, using equation (3.3), by

$$\sigma = G^*(j\omega)\frac{\partial\xi}{\partial y} = G^*(j\omega)[-j\Gamma\xi] = -G^*(j\omega)\left[j\omega\xi\sqrt{\frac{\rho}{G^*(j\omega)}}\right]$$

$$= -j\omega\xi\sqrt{\rho G^*(j\omega)}$$

Thus
$$Z_L = \frac{j\omega\xi\sqrt{\rho G^*(j\omega)}}{j\omega\xi} = \sqrt{\rho G^*(j\omega)}$$

or
$$Z_L^2 = \rho G^*(j\omega) \tag{3.5}$$

For a Newtonian liquid, where $G^*(j\omega) = j\omega\eta$, the components R_L and X_L of the complex impedance are given by

$$Z_L = R_L + jX_L = (1+j)\sqrt{\pi f\eta\rho} \tag{3.6}$$

i.e. R_L is equal in magnitude to X_L. It follows from equation (3.5) that the components of the complex modulus, $G^*(j\omega) = G^*(\omega) + jG''(\omega)$, are related to the components of the impedance by equations (3.7) and (3.8).

$$G'(\omega) = \frac{R_L^2 - X_L^2}{\rho} \tag{3.7}$$

$$G''(\omega) = \frac{2R_L X_L}{\rho} \tag{3.8}$$

In general the shear modulus will not have the simple form given above for a Newtonian liquid except at low frequencies where sufficient time is available during each stress cycle for viscous flow to occur. At higher frequencies the time required for molecular translation becomes comparable with the period of the stress cycle and the liquid exhibits a shear rigidity. At sufficiently high frequencies the behaviour will be purely elastic, with no molecular translation occurring during each cycle, and consequently negligible energy loss due to viscous flow. Under these conditions the liquid behaves as a glass, although the temperature may be well above the conventional glass transition temperature T_g.

The real component of the complex modulus, $G'(\omega)$, the ratio of the stress in phase with the strain to the strain, is termed the storage modulus, because of the association with the storage and release of elastic energy during the periodic deformation. The imaginary component, $G''(\omega)$, the ratio of the stress 90° out of phase with the strain to the strain, is called the loss modulus, because of the association with the dissipation of energy as heat by viscous flow. The modulus components are subject, for a liquid, to the following limiting conditions.

At low frequencies the behaviour is purely viscous, or Newtonian.

Then
$$\lim_{\omega \to 0} G'(\omega) = 0$$

$$\lim_{\omega \to 0} G''(\omega) = \omega \eta$$

At high frequencies the behaviour is purely elastic:

$$\lim_{\omega \to \infty} G'(\omega) = G_\infty$$

$$\lim_{\omega \to \infty} G''(\omega) = 0$$

G_∞ is the limiting elastic modulus, or the instantaneous elastic modulus.

The liquid properties may be alternatively represented by a complex viscosity defined by

$$\eta^*(j\omega) = \eta'(\omega) - j\eta''(\omega) = \frac{G^*(j\omega)}{j\omega}. \quad (3.9)$$

Then $\quad \eta'(\omega) = G''(\omega)/\omega$

and $\quad \eta''(\omega) = G'(\omega)/\omega$.

The low frequency limit of $\eta'(\omega)$ is $\omega\eta/\omega = \eta$. Thus $\eta'(\omega)$, the dynamic viscosity, becomes equal to the steady flow Newtonian viscosity in the low frequency limit. The complex compliance, $J^*(j\omega)$, is the inverse of the complex modulus, so that

$$J^*(j\omega) = J'(\omega) - jJ''(\omega) = 1/G^*(j\omega). \quad (3.10)$$

The real and imaginary components, $J'(\omega)$ and $J''(\omega)$ are called the storage compliance and loss compliance, respectively.

The inverse of the complex viscosity is the complex fluidity, $\mu^*(j\omega)$, so that $\mu^*(j\omega) = \mu'(\omega) + j\mu''(\omega) = 1/\eta^*(j\omega)$.

Table 3.1 summarizes the above viscoelastic functions and their interrelations, and gives the main limiting values of these functions.

TABLE 3.1 Definitions and relations between viscoelastic functions

shear stress = σ, acting in x direction, on x-z plane
displacement in x direction = ξ
shear strain = $\gamma = \partial\xi/\partial y$
shear rate, rate of strain = $\dot{\gamma} = \partial\gamma/\partial t = \partial\dot{\xi}/\partial y$
particle velocity in x direction = $\dot{\xi} = \partial\xi/\partial t$

Viscoelastic functions, harmonic variation of stress and strain of angular frequency $\omega = 2\pi f$.

σ/γ = complex shear modulus   = $G^*(j\omega) = G'(\omega) + jG''(\omega)$
$\sigma/\dot{\gamma}$ = complex viscosity   = $\eta^*(j\omega) = \eta'(\omega) - j\eta''(\omega)$
γ/σ = complex shear compliance = $J^*(j\omega) = J'(\omega) - jJ''(\omega)$
$\dot{\gamma}/\sigma$ = complex fluidity   = $\mu^*(j\omega) = \mu'(\omega) + j\mu''(\omega)$
$-\sigma/\dot{\xi}$ = complex mechanical impedance = Z_L^* = $R_L + jX_L$

$J^*(j\omega) = 1/G^*(j\omega) = \mu^*(j\omega)/j\omega = 1/j\omega\eta^*(j\omega) = \rho/Z_L^2$

Viscoelastic functions, transient variation of stress and strain.

σ/γ_0 = relaxation modulus = $G(t)$
      = response after step function of strain γ_0
γ/σ_0 = creep compliance = $J(t)$
      = response after step function of stress σ_0

Limiting values of viscoelastic functions

$\lim_{\omega \to \infty} G'(\omega) = \lim_{t \to 0} G(t) = G_\infty$, limiting high frequency shear modulus (glassy modulus, instantaneous modulus)

$\lim_{\omega \to \infty} J'(\omega) = \lim_{t \to 0} J(t) = J_\infty = 1/G_\infty$, limiting high frequency compliance (glassy compliance, instantaneous compliance)

$\lim_{\omega \to 0} J'(\omega) = \lim_{t \to \infty} J(t) - t/\eta = J_e$, equilibrium compliance

$\lim_{\omega \to 0} \eta'(\omega) = \lim_{\omega \to 0} G''(\omega)/\omega = \eta$, steady flow (Newtonian) viscosity

3.2 The Maxwell Element

It is often convenient to visualize the behaviour of a complex material in terms of models. Historically, mechanical models have been used extensively, although electrical analogues could equally well be used. The basic mechanical model elements are a coiled spring to represent Hookean elastic deformation, and a dashpot — a piston operating in a cylinder of a Newtonian liquid — to represent Newtonian viscous flow. Extension of the elements is analogous to shear strain, the associated force being analogous to the shear stress.

The combination of a spring in series with a dashpot was studied by Maxwell (1886). This simple model, shown in Fig. 3.2, exhibits both viscous and elastic behaviour. Viscous flow in the dashpot with negligible extension of the spring takes place if the extension rate is small. If the model is rapidly extended and immediately released, the deformation is purely elastic, as insufficient time is available for flow to occur in the dashpot. Between these extremes, the behaviour will be a combination of both elastic and viscous.

FIG. 3.2. Maxwell element. The spring corresponds to a shear modulus G, the dashpot corresponds to a viscosity η.

The basic equations for the components of the model are:

for the dashpot $\qquad \sigma = \eta \dot{\gamma}_N$

for the spring $\qquad \sigma = G \gamma_H$

where σ is the applied stress, $\dot{\gamma}_N$ is the rate of extension of the dashpot and γ_H is the extension of the spring. The rate of extension of the spring is $\dot{\gamma}_H = \dot{\sigma}/G$, the total rate of extension is then

$$\dot{\gamma} = \dot{\gamma}_N + \dot{\gamma}_H$$

$$= \frac{\sigma}{\eta} + \frac{\dot{\sigma}}{G}$$

or

$$\sigma + \frac{\eta}{G} \dot{\sigma} = \eta \dot{\gamma} \qquad (3.11)$$

This is the constitutive equation of the Maxwell element. The ratio η/G has the dimensions of a time and is called the Maxwell relaxation time τ_m, i.e.,

$$\tau_m = \eta/G \qquad (3.12)$$

3.2.1 OSCILLATORY RESPONSE OF THE MAXWELL ELEMENT

For sinusoidal variations of stress and strain of frequency ω, equation (3.11) becomes

$$\sigma + j\omega\tau_m \sigma = j\omega\eta\gamma$$

so that

$$G^*(j\omega) = \frac{\sigma}{\gamma} = \frac{j\omega\eta}{1 + j\omega\tau_m} \qquad (3.13)$$

Rationalising this expresssion

$$G^*(j\omega) = \frac{\omega^2 \eta \tau_m + j\omega\eta}{1 + \omega^2 \tau_m^2}$$

But $\eta = G\tau_m$ so that $G^*(j\omega) = G \dfrac{\omega^2 \tau_m^2 + j\omega\tau_m}{1 + \omega^2 \tau_m^2}$

Then

$$G'(\omega) = G \frac{\omega^2 \tau_m^2}{1 + \omega^2 \tau_m^2}$$

which reduces to $G'(\omega) = G$ in the limit as $\omega \to \infty$; but this limiting value has been defined as G_∞. Thus the spring in the Maxwell element corresponds to the instantaneous, or limiting high frequency modulus of a liquid. The loss modulus is given by

$$G''(\omega) = G_\infty \frac{\omega\tau_m}{1 + \omega^2 \tau_m^2}$$

which in the limit as $\omega \to 0$ becomes $G''(\omega) = G_\infty \omega\tau_m = \omega\eta$. The dashpot of the Maxwell element therefore corresponds to the steady flow viscosity of a liquid. In normalised form, the variation with frequency of the modulus components and the dynamic viscosity is given by equation (3.14).

$$\frac{G'(\omega)}{G_\infty} = \frac{\omega^2 \tau_m^2}{1 + \omega^2 \tau_m^2}$$

$$\frac{G''(\omega)}{G_\infty} = \frac{\omega\tau_m}{1 + \omega^2 \tau_m^2} \qquad (3.14)$$

3. LINEAR VISCOELASTICITY

$$\frac{\eta'(\omega)}{\eta} = \frac{G''(\omega)}{\omega\eta} = \frac{1}{1+\omega^2\tau_m^2}$$

The complex compliance $J^*(j\omega)$ for the Maxwell element is given by the simple expression

$$J^*(j\omega) = \frac{\gamma}{\sigma} = \frac{1+j\omega\tau_m}{j\omega\eta} = \frac{1}{G_\infty} - j\frac{1}{\omega\eta} \qquad (3.15)$$

Figure 3.3 shows the frequency variation of the modulus components and the dynamic viscosity for the Maxwell element; Fig. 3.4 shows the corresponding variation of the components of the complex shear impedance, together with the variation for the case of a Newtonian liquid.

3.2.2 TRANSIENT RESPONSE OF THE MAXWELL ELEMENT

The shear stress in a liquid subjected to a step function of applied shear strain, rises immediately to a value determined by the instantaneous elastic modulus, and then decays steadily as the liquid flows and takes up a new configuration. The Maxwell element

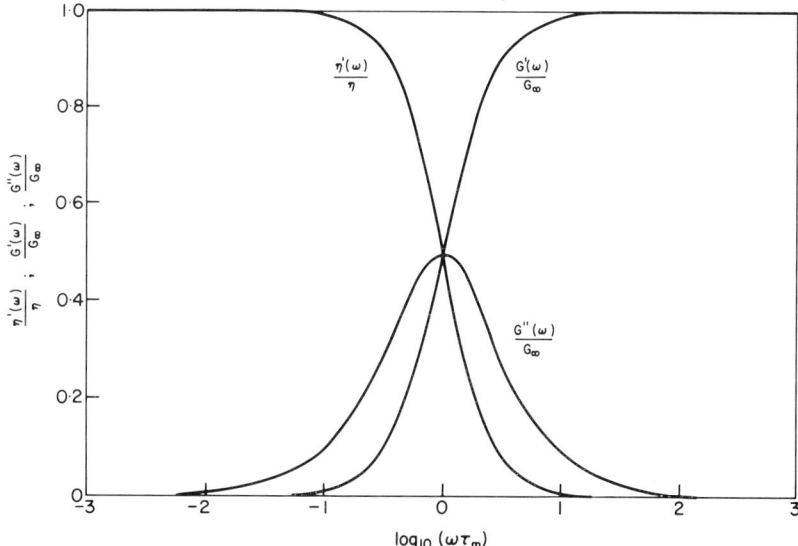

FIG. 3.3. Maxwell element: variation of the normalised components of the shear modulus and the normalised dynamic viscosity as a function of normalised frequency.

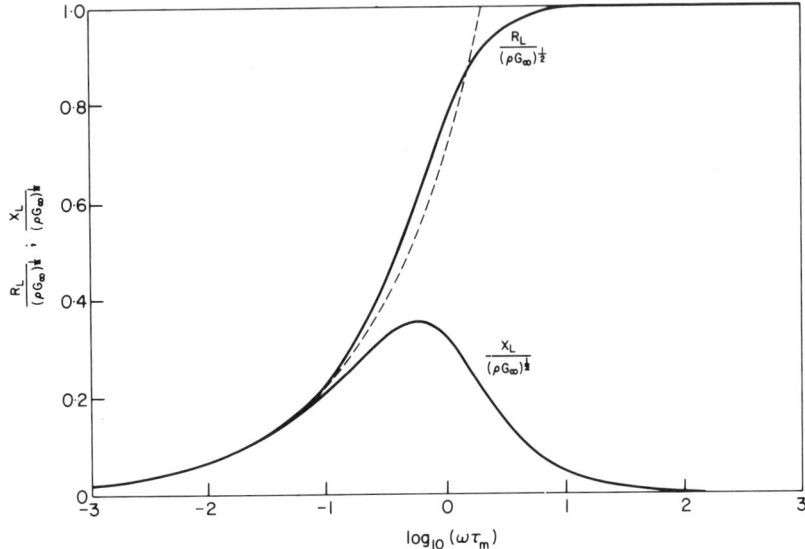

FIG. 3.4. Maxwell element: variation of the normalised components of the shear impedance as a function of normalised frequency. The dashed curve shows the variation of both components for a Newtonian liquid ($R_L = X_L$).

exhibits this form of "stress relaxation" behaviour, the variation of stress with time being given by integrating equation (3.11). Then

$$\sigma = \sigma_0 \exp[-t/\tau_m]$$

where the initial value of the stress $\sigma_0 = G_\infty \gamma_0$, γ_0 being the magnitude of the applied step function of strain. The stress relaxation modulus $G(t)$ is given by

$$G(t) = \frac{\sigma}{\gamma_0} = G_\infty \exp[-t/\tau_m] \qquad (3.16)$$

The stress decays exponentially with time, the Maxwell relaxation time τ_m being the time at which the stress is reduced to $1/e$ of its original value, as shown in Fig. 3.5.

The converse experiment, in which a step function of stress is applied and the variation of strain with time is observed, is known as creep. The response in this case is

$$\gamma = \sigma_0 \frac{t}{\eta} + \frac{\sigma_0}{G_\infty}$$

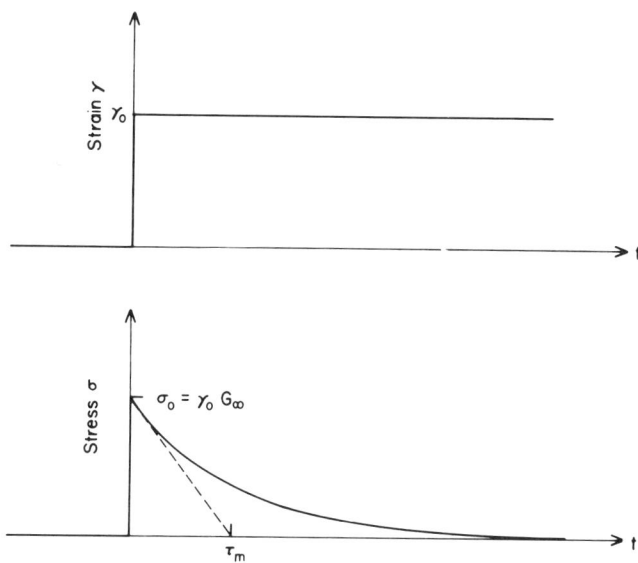

FIG. 3.5. Stress relaxation behaviour of the Maxwell element.

where σ_0 is the magnitude of the step function of stress. The creep compliance $J(t)$ is

$$J(t) = \frac{\gamma}{\sigma_0} = \frac{1}{G_\infty} + \frac{t}{\eta} \qquad (3.17)$$

If the stress is removed at time $t = T$, the elastic deformation σ_0/G_∞ is immediately recovered, the final strain $(\sigma_0 T/\eta)$ being due entirely to viscous flow. The creep response of the Maxwell element is shown in Fig. 3.6.

3.3 The Voigt Element

The immediate attainment of steady viscous flow which is characteristic of the creep response of the Maxwell element is not generally found in liquids, and in this respect the Maxwell element is an unsatisfactory liquid model. Steady viscous flow is attained gradually after the application of stress (neglecting inertia effects) with additional elastic energy being stored in the liquid. This retarded elastic behaviour is shown by the combination of a spring and

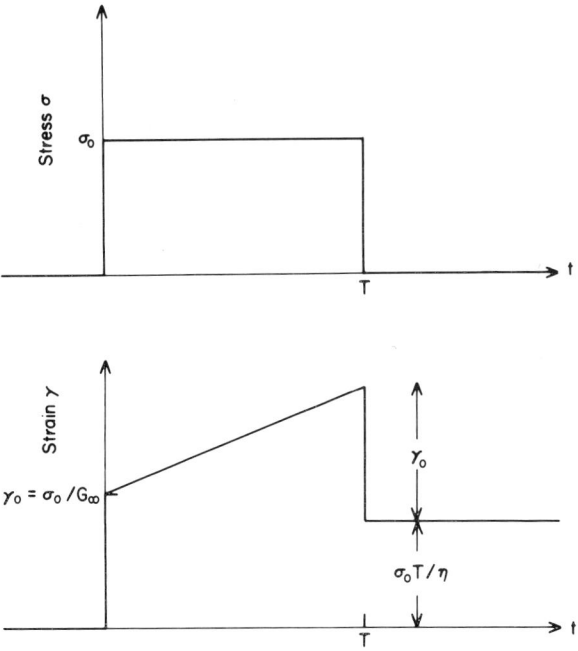

FIG. 3.6. Creep behaviour of Maxwell element.

dashpot in parallel shown in Fig. 3.7, an arrangement usually called the Voigt, or Kelvin element (Reiner, 1958, p. 457). The Voigt element alone is unsatisfactory as a liquid model, however, as steady viscous flow is not possible, but it is used as a component of more complicated model systems.

Using the same basic equations as for the Maxwell element, but

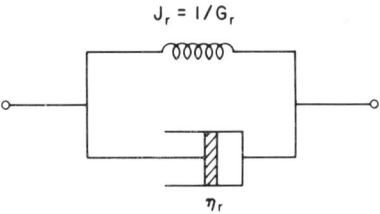

FIG. 3.7. The Voigt element. The spring corresponds to a shear modulus G_r, the dashpot corresponds to a viscosity η_r.

here adding the stresses and taking the deformation as the common parameter, gives the constitutive equation of the Voigt element as

$$\sigma = \sigma_H + \sigma_N$$

$$= \eta_r \dot{\gamma} + \frac{1}{J_r} \gamma \qquad (3.18)$$

where η_r is the viscosity associated with the dashpot, and $J_r (= 1/G_r)$ is the compliance associated with the spring.

3.3.1 OSCILLATORY RESPONSE OF THE VOIGT ELEMENT

For sinusoidal variations of stress and strain, equation (3.18) becomes

$$\sigma = j\omega \eta_r \gamma + \frac{\gamma}{J_r}$$

giving a complex compliance $J^*(j\omega)$ of the form

$$J^*(j\omega) = \frac{\gamma}{\sigma} = J_r \frac{1}{1 + j\omega \eta_r J_r}$$

or
$$J^*(j\omega) = J_r \frac{1}{1 + j\omega \tau_r} \qquad (3.19)$$

where $\tau_r = \eta_r J_r$ is a retardation time. The components of $J^*(j\omega)$ are then

$$J'(\omega) = J_r \frac{1}{1 + \omega^2 \tau_r^2}$$

$$J''(\omega) = J_r \frac{\omega \tau_r}{1 + \omega^2 \tau_r^2} \qquad (3.20)$$

The limiting value at low frequencies of the storage modulus is

$$\lim_{\omega \to 0} J'(\omega) = J_r$$

Thus J_r is an equilibrium or steady state compliance. The limiting value at high frequencies is zero, as no instantaneous deformation of the Voigt element is possible. The variation of the components $J'(\omega)$ and $J''(\omega)$ with frequency is shown in Fig. 3.8.

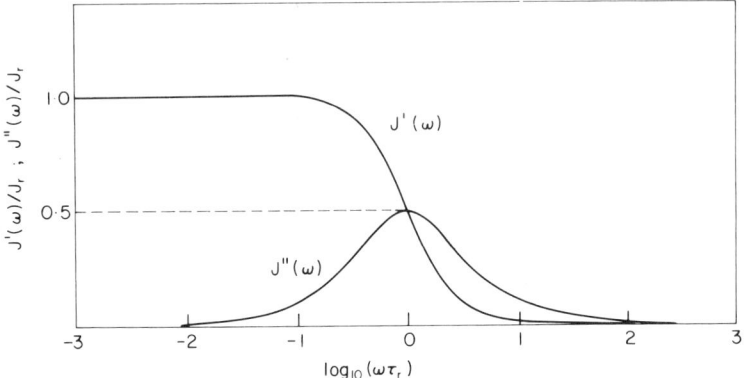

FIG. 3.8. Voigt element: variation of the normalised components of the shear compliance as a function of normalised frequency.

3.3.2 TRANSIENT RESPONSE OF THE VOIGT ELEMENT

The utility of the Voigt element lies mainly in its response to a step function of stress — the creep response. Integrating equation (3.18) gives the time variation of strain, after an applied step function of stress of magnitude of σ_0 as

$$\gamma = \sigma_0 J_r (1 - e^{-t/\tau_r})$$

The creep compliance is then

$$J(t) = \frac{\gamma}{\sigma_0} = J_r(1 - e^{-t/\tau_r}) \qquad (3.21)$$

The opposite behaviour, with an exponential decay of strain, occurs if the stress is removed, as shown in Fig. 3.9. The retardation time τ_r is the time taken for the strain to reach $(1 - 1/e)$ of the equilibrium value of $\sigma_0 J_r$. This slow approach to an equilibrium elastic strain is called retarded elasticity, and J_r is alternatively called a retarded elastic compliance.

3.4 Viscoelastic Models for Complex Systems

Both the Maxwell element and the Voigt element suffer from serious deficiencies when their responses are compared with the response of a viscoelastic liquid. Particularly, the Maxwell element exhibits an

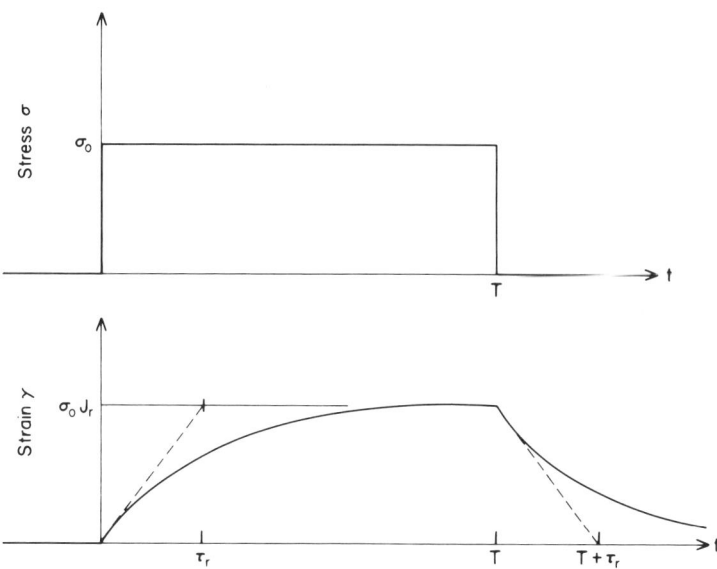

FIG. 3.9. Creep response of Voigt element.

unrealistic creep response, and the Voigt element allows for neither viscous flow or any instantaneous deformation. A combination of these two elements gives a model, sometimes referred to as the Burgers model (Reiner, 1958, p. 472) which does show the main features of liquid behaviour. This model is shown in Fig. 3.10.

This system exhibits both an instantaneous and a retarded elastic response, as well as steady viscous flow. When steady state conditions are reached after application of a constant stress σ_0, and steady viscous flow characterised by the viscosity η is established, the total elastic deformation is given by the steady state compliance $J_e = J_\infty + J_r$. If then after time T the stress is removed, this stored elastic strain

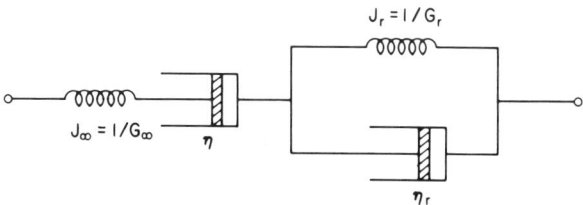

FIG. 3.10. Burgers model system.

is gradually recovered — a process known as creep recovery — and the final strain is given by $\sigma_0 T/\eta$. This response is shown in Fig.3.11, and the equations governing the creep and oscillatory responses are equations (3.22) and 3.23), obtained by summing the responses of the separate Maxwell and Voigt elements.

$$J(t) = J_\infty + \frac{t}{\eta} + J_r(1 - e^{-t/\tau_r}) \qquad (3.22)$$

$$J^*(j\omega) = J_\infty + \frac{1}{j\omega\eta} + J_r \frac{1}{1 + j\omega\tau_r} \qquad (3.23)$$

The retardational contribution to $J^*(j\omega)$ is denoted by $J_r^*(j\omega)$ so that

$$J^*(j\omega) = J_\infty + \frac{1}{j\omega\eta} + J_r^*(j\omega)$$

and $\qquad J_r^*(j\omega) = J_r'(\omega) - jJ_r''(\omega) = J_r \dfrac{1}{1 + j\omega\tau_r}$.

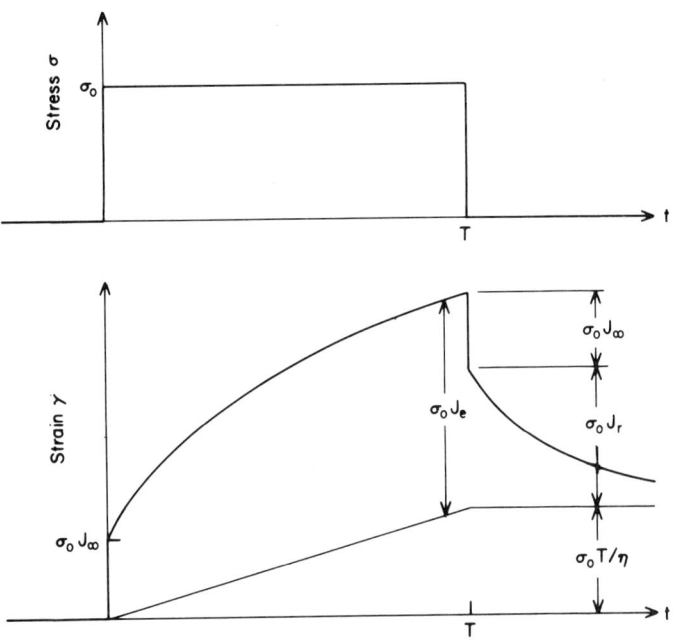

FIG. 3.11. Creep response of Burgers model.

The variation of the components of $J_r^*(j\omega)$ and $G^*(j\omega)$ are shown in Fig. 3.12. The curves have been calculated for a compliance ratio J_r/J_∞ of 10, and for values of the ratio τ_r/τ_m of 1, 10 and 100. It is apparent that for values of $\tau_r/\tau_m > 10$, as is found to be the case in many supercooled liquids, the Maxwell process provides the major contribution to the total shear modulus, and the retardational

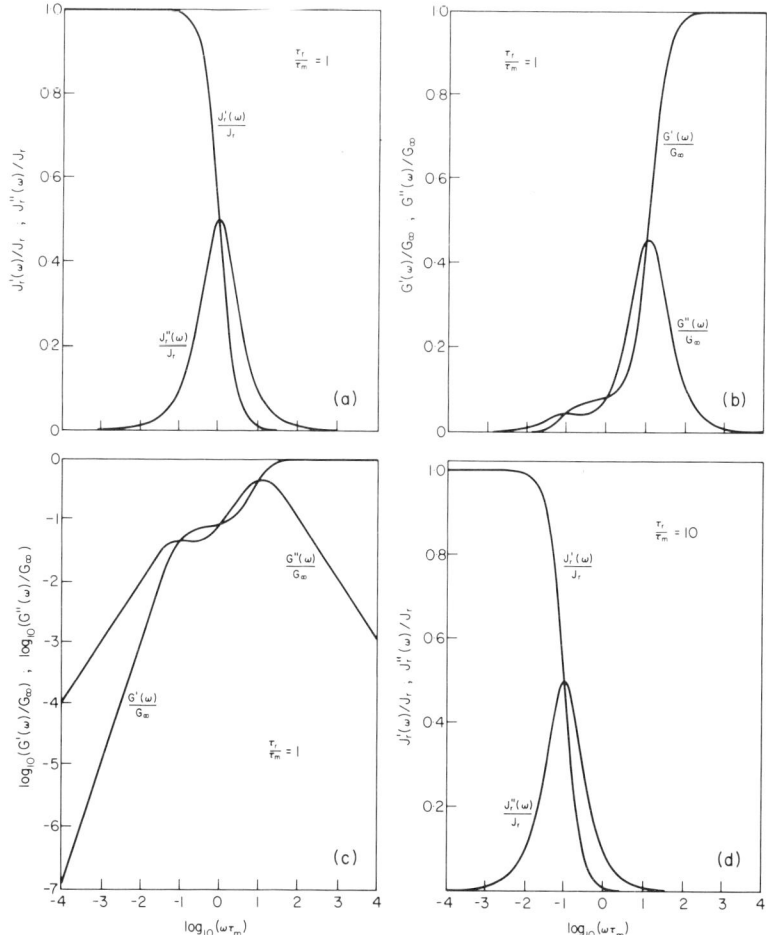

FIG. 3.12. Variation of the normalized components of $J_r^*(j\omega)$ and $G^*(j\omega)$ with frequency for the Burgers model (equation 3.23). The curves have been calculated for a ratio $J_r/J_\infty = 10$ and for values of the ratio τ_r/τ_m of 1, 10 and 100. (*Continued overleaf*).

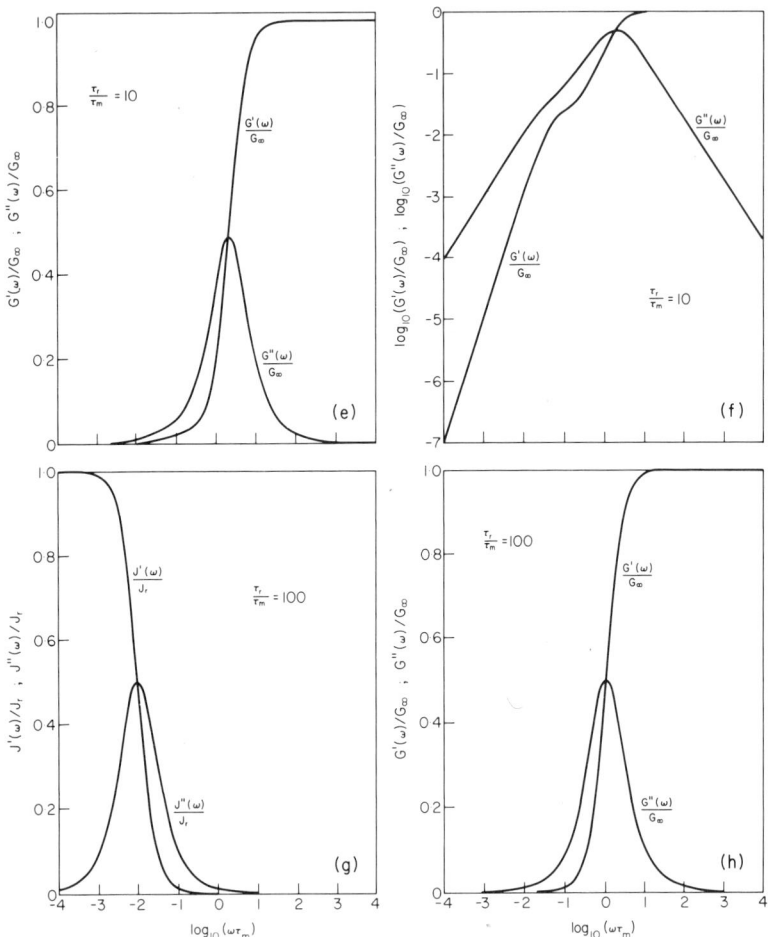

FIG. 3.12. (*Cont.*) Variation of the normalized components of $J_r^*(j\omega)$ and $G^*(j\omega)$ with frequency for the Burgers model (equation 3.23). The curves have been calculated for a ratio $J_r/J_\infty = 10$ and for values of the ratio τ_r/τ_m of 1, 10 and 100. (*Continued on next page*).

contribution is only significant at frequencies below $\omega \tau_m = 1$ where the value of the modulus is small.

Although the response of the Burgers model comes nearer to representing the behaviour of a real liquid than either the Maxwell or Voigt elements alone, the model response is still a drastic simplification of real behaviour. The stress relaxation of the Maxwell

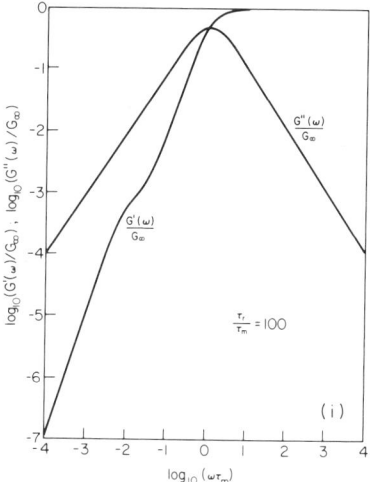

FIG. 3.12. (*Cont.*) Variation of the normalized components of $J_r^*(j\omega)$ and $G^*(j\omega)$ with frequency for the Burgers model (equation 3.23). The curves have been calculated for a ratio $J_r/J_\infty = 10$ and for values of the ratio τ_r/τ_m of 1, 10 and 100.

element and the creep of the Voigt element are exponential functions of time, whereas most observed creep and stress relaxation processes progress more gradually than the predictions of the simple models. Also, the change in the values of the components of the complex modulus, or compliance, with frequency are generally less rapid than the single relaxation or retardation time processes of the models. A closer approach to real material behaviour may be achieved by combining a number of basic model elements together, and adjusting the individual parameters until the response follows the experimentally determined behaviour to the desired degree of accuracy. The choice of whether to use a number of Maxwell elements in parallel, a number of Voigt elements in series, or some combination of the two types is purely arbitrary. The same final result can be achieved in all cases. The interrelations of such "canonic" types of model element are well understood, using analogies with electrical filter network theories, and a considerable literature exists on the subject (Gross, 1953; Alfrey, 1945; Whitehead, 1944; Roscoe, 1950; Guillemin, 1947).

It is mathematically convenient to describe a group of basic model elements in terms of a "spectrum" of relaxation or retardation times, but it should be noted that such a spectrum remains a purely empirical way of describing experimental results, and it does not relate directly to any physical process in a liquid.

3.4.1 THE RELAXATION SPECTRUM

A generalized form of Maxwell model consists of an arbitrary number of Maxwell elements in parallel, each having a viscosity η_i and a rigidity modulus G_i, where the subscript i numbers the elements. The overall response of the system is then given by summing equations (3.14) and (3.16) over all the elements since the stresses are additive. Then if there are a total of n elements,

$$G(t) = \sum_{i=1}^{n} G_i e^{-t/\tau_i} \qquad (3.24)$$

$$G^*(j\omega) = \sum_{i=1}^{n} G_i \frac{j\omega\tau_i}{1 + j\omega\tau_i} \qquad (3.25)$$

where $\tau_i = \eta_i/G_i$. In this way the system is represented by a spectrum of discrete relaxation processes, each having a strength given by the modulus G_i and associated with a time τ_i. The resolution, from experimental data of limited accuracy, of the individual values of G_i and τ_i for a number of discrete processes is difficult, however, and it is usually more convenient to represent the behaviour in terms of a continuous spectrum of relaxation processes. Each infinitesimal contribution $H(\tau)d\tau$ to the rigidity is associated with times lying in the range τ to $\tau + d\tau$. The distribution function $H(\tau)$ is called the relaxation spectrum. In terms of $H(\tau)$, the behaviour of the generalised Maxwell system is expressed as follows.

$$G(t) = \int_0^{+\infty} H(\tau) e^{-t/\tau} \, d\tau \qquad (3.26)$$

$$G^*(j\omega) = \int_0^{+\infty} H(\tau) \frac{j\omega\tau}{1 + j\omega\tau} \, d\tau \qquad (3.27)$$

or, giving the separate components,

$$G'(\omega) = \int_0^{+\infty} H(\tau) \frac{\omega^2 \tau^2}{1 + \omega^2 \tau^2} \, d\tau \qquad (3.28)$$

$$G''(\omega) = \int_0^{+\infty} H(\tau) \frac{\omega \tau}{1 + \omega^2 \tau^2} \, d\tau \qquad (3.29)$$

$$\eta'(\omega) = \int_0^{+\infty} H(\tau) \frac{1}{1 + \omega^2 \tau^2} \, d\tau \qquad (3.30)$$

The limiting quantities G_∞ and η may also be expressed in terms of $H(\tau)$. Putting $\omega = \infty$ in equation (3.28) gives G_∞:

$$G_\infty = \int_0^{+\infty} H(\tau) d\tau \qquad (3.31)$$

This implies that for a finite value of G_∞, $H(\tau)$ must vanish at both short and long times. The steady flow viscosity is given by putting $\omega = 0$ in equation 3.30:

$$\eta = \int_0^{+\infty} \tau H(\tau) d\tau \qquad (3.32)$$

3.4.2 THE RETARDATION SPECTRUM

A generalised Voigt model comprises a number of Voigt elements connected in series. To represent a liquid, one element must have only a viscous component ($J_r = \infty$), to allow for steady state viscous flow, and a second element must have zero viscosity so that instantaneous deformation is possible. The remaining elements can then be summed to give a discrete spectrum of retardation processes, or in the limit, regarded as a continuous distribution of processes. Then the infinitesimal contribution $L(\tau)d\tau$ to the compliance is associated with retardation times lying between τ and $\tau + d\tau$; $L(\tau)$ is the retardation spectrum. The behaviour is then expressed as follows:

$$J(t) = J_\infty + \frac{t}{\eta} + \int_0^{+\infty} L(\tau)(1 - e^{-t/\tau}) d\tau \qquad (3.33)$$

$$J^*(j\omega) = J_\infty + \frac{1}{j\omega\eta} + \int_0^{+\infty} L(\tau) \frac{1}{1 + j\omega\tau} d\tau \qquad (3.34)$$

or in terms of the components,

$$J'(\omega) = J_\infty + \int_0^{+\infty} L(\tau) \frac{1}{1+\omega^2\tau^2} d\tau \qquad (3.35)$$

$$J''(\omega) = \frac{1}{\omega\eta} + \int_0^{+\infty} L(\tau) \frac{\omega\tau}{1+\omega^2\tau^2} d\tau \qquad (3.36)$$

The limiting value of $J(t)$ as $t \to \infty$, minus the steady flow term, gives the steady state compliance J_e.

$$\lim_{t \to \infty} \left(J(t) - \frac{t}{\eta} \right) = J_e = J_\infty + \int_0^{+\infty} L(\tau) d\tau \qquad (3.37)$$

The limiting value of $J'(\omega)$ as $\omega \to 0$ also has the same value:

$$\lim_{\omega \to 0} J'(\omega) = J_\infty + \int_0^{+\infty} L(\tau) d\tau \qquad (3.38)$$

The equilibrium value of the retarded compliance J_r is then given by

$$J_r = \int_0^{+\infty} L(\tau) d\tau \qquad (3.39)$$

If $J_r^*(j\omega)$ is the retardational contribution to $J^*(j\omega)$ so that

$$J^*(j\omega) = J_\infty + 1/j\omega\eta + J_r^*(j\omega)$$

then the components of $J_r^*(j\omega) = J_r'(\omega) - jJ_r''(\omega)$ are given by

$$J_r'(\omega) = J'(\omega) - J_\infty = \int_0^{+\infty} L(\tau) \frac{1}{1+\omega^2\tau^2} d\tau \qquad (3.40)$$

$$J_r''(\omega) = J''(\omega) - \frac{1}{\omega\eta} = \int_0^{+\infty} L(\tau) \frac{\omega\tau}{1+\omega^2\tau^2} d\tau \qquad (3.41)$$

The subtraction of the "Maxwell" terms from $J^*(j\omega)$ to leave $J^*_r(j\omega)$ is found to facilitate comparisons between the behaviour of different liquids; this is a reason for the use of the compliance instead of the complex modulus $G^*(j\omega)$, where no such separation of the contributions to the overall behaviour is possible.

The steady state compliance J_e may also be determined from the low frequency region of $G'(\omega)$. In the limit as ω tends towards zero, $J'(\omega) \to J_e$ and $J''(\omega) \to 1/\omega\eta$. Then from equation (3.10),

$$G'(\omega) = \frac{J'(\omega)}{[J'(\omega)]^2 + [J''(\omega)]^2}$$

$$\lim_{\omega \to 0} G'(\omega) = \omega^2 \eta^2 J_e \qquad (3.42)$$

Because the time-scale often extends over several orders of magnitude, viscoelastic functions are commonly expressed as a function of a logarithmic time scale. Then, in terms of a normalised time $\tau' = \tau/\tau_0$, where τ_0 is an arbitrary normalising time, a relaxation spectrum $H(\ln\tau')$ is defined such that each infinitesimal contribution $H(\ln\tau')d(\ln\tau')$ to the rigidity is associated with times lying in the range $\ln\tau'$ to $\ln\tau' + d(\ln\tau')$. Then, for example, equation (3.31) becomes

$$G_\infty = \int_{-\infty}^{+\infty} H(\ln\tau')d(\ln\tau')$$

where $H(\ln\tau') = \tau H(\tau)$. The shape of the spectrum is thus considerably altered by the change of the variable. The units of $H(\ln\tau')$ are those of a modulus, i.e. Pa, whereas $H(\tau)$ has units of Pa s^{-1}.

By a similar argument, a retardation function $L(\ln\tau')$ may be defined, such that $L(\ln\tau') = \tau L(\tau)$ and for example, equation (3.39) becomes

$$J_r = \int_{-\infty}^{+\infty} L(\ln\tau')d(\ln\tau').$$

The need to use a normalised time τ', so that the argument of the logarithm is dimensionless, is often not stated explicitly, and variables such as $\ln\tau$ are often used. In such cases a normalisation time of unity is implied.

3.5 Interrelations Between Viscoelastic Functions

The interrelations between the many viscoelastic functions are treated in detail in the literature on the theories of linear viscoelasticity. In principle, if any one function is known over the entire time, or frequency range, then any other function can be evaluated. Difficulties arise in practice, however, from the limited time or frequency range of most experimental data, and from the limited accuracy of the data. In the usual case where analytic forms for the functions are not known, approximate numerical methods must be used. Such methods, widely used in the analysis of the behaviour of polymers, are discussed by Ferry (1970), Schwarzl (1969) and Schwarzl and Struik (1967). The formal interrelations between the functions are shown in Fig. 3.13, taken from Gross (1953).

3.6 Method of Reduced Variables

The transition from purely viscous to purely elastic behaviour is found to extend over many decades of frequency (or time); for simple supercooled liquids, viscoelastic behaviour is observed for about 3 decades — see Chapter 5 — but for polymer solutions and melts the width of the viscoelastic region may be much wider, extending in some cases to over 15 decades (Tobolsky 1958). It is generally not possible to cover this wide range experimentally, and the position of the viscoelastic region on the time scale also depends on the temperature of measurement. This fact is utilized to bring the region of interest within the range of a particular apparatus, and the use of temperature as an additional variable to frequency, or time, often enables the complete region of viscoelastic behaviour to be explored with an apparatus of limited frequency range. The process of reducing data obtained over a range of temperatures to the behaviour which would be observed at a single temperature, but over a much wider frequency, or time, range is usually referred to as the method of reduced variables. Other names for this principle which have been used are time-temperature superposition, time-temperature reducibility, and thermorheological simplicity. The resulting plots using reduced variables are often called master curves.

It has been implied, in the previous analysis, that viscoelastic behaviour is a function of frequency or time only, and the viscoelastic region may be explored by varying ω in the dimensionless parameter $\omega\tau$ (or t in the parameter t/τ). The region could also be explored by varying the time constant τ, keeping ω or t constant, however, if the time constant could be varied in a

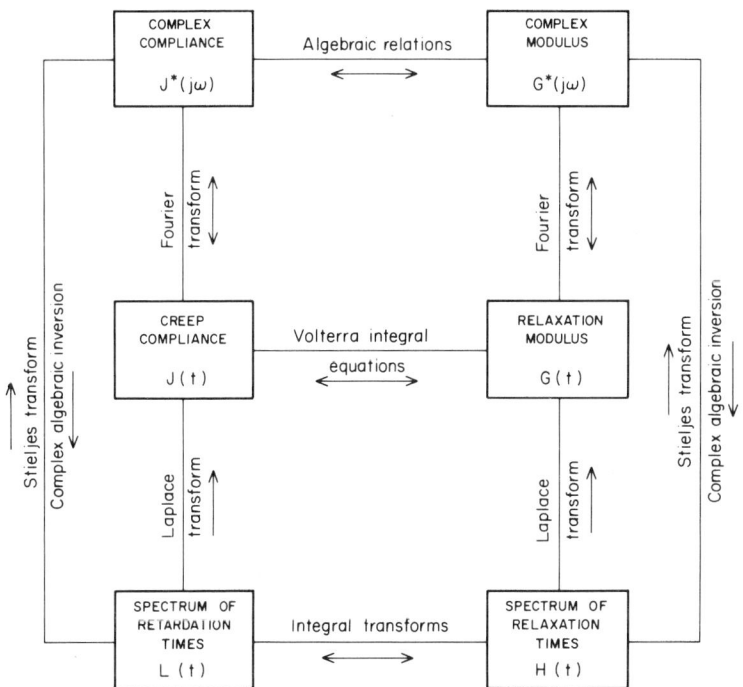

FIG. 3.13. Relations between the viscoelastic functions (after Gross, 1953).

controlled manner. Taking the Maxwell model as a simple example, the relaxation time τ_m is dependent on the viscosity and the modulus G_∞. Both of these quantities vary in magnitude with temperature, G_∞ in an approximately linear manner, the viscosity in an approximately exponential manner. The variation of τ_m with temperature is thus strongly linked to the variation of viscosity with temperature, and temperature can be used as a variable with which to explore the relaxation region. Thus at low temperatures, the viscosity is high, τ_m has a high value and $\omega\tau_m \gg 1$: thus the behaviour will be elastic. At high temperatures, the viscosity and the relaxation time have low values, $\omega\tau_m \ll 1$ and the behaviour is viscous. A striking example in the use of temperature to explore viscoelastic behaviour covering some 10 decades of reduced frequency, from an experimental range of only 2 decades, is shown in Figs. 3.14 and 3.15, taken from the results of Dannhauser et al. (1958).

In order to plot data in this manner using reduced variables, ideally the variation of τ_m and G_∞ with temperature must be known. However, it is possible to reduce data empirically, merely by shifting the curves until a smooth master curve is obtained. When this method is used, either because viscosity and G_∞ data are not available, or because it does not apply to the viscoelastic region being studied, as can be the case for the glassy region of polymer melts,

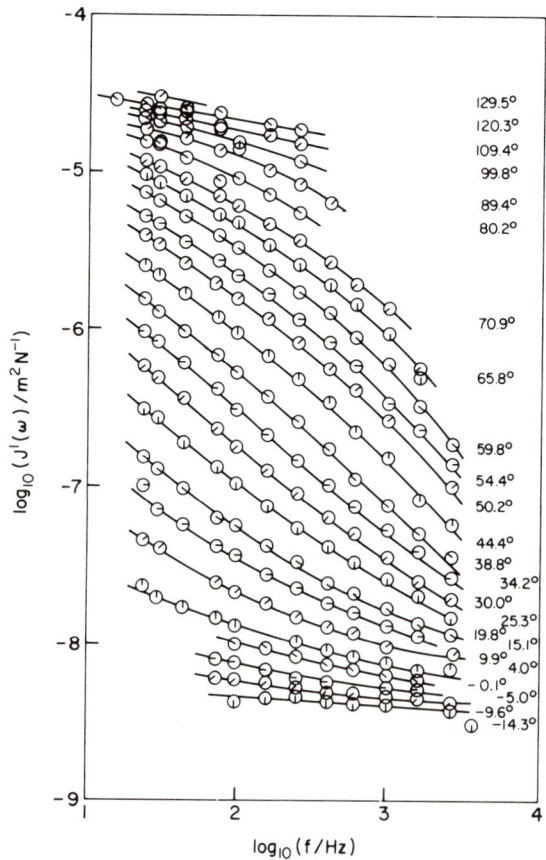

FIG. 3.14. Variation of $J'(\omega)$ as a function of frequency at various temperatures for poly (n-octyl)methacrylate (after' Dannhauser, *et al.* 1958).

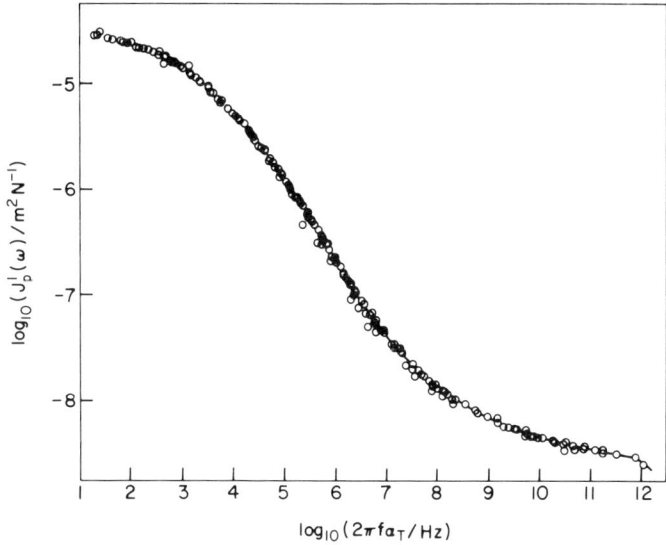

FIG. 3.15. Data of figure 3.14 plotted using reduced variables to obtain a "master curve" referred to a temperature of 100°C (after Dannhauser, et al. 1958).

care must be taken to avoid obtaining erroneous results. This topic is discussed in detail by Ferry (1970, Chapter 11).

The temperature dependence of τ_m and G_∞ may be specified by the variables α_T and β_T which are functions of temperature. Then

$$[G_\infty]_T = \beta_T [G_\infty]_0$$
and
$$[\tau_m]_T = \alpha_T [\tau_m]_0 \quad (3.43)$$

where $[G_\infty]_T$ and $[\tau_m]_T$ are the values at the temperature of measurement T, and $[G_\infty]_0$ and $[\tau_m]_0$ are the values at an arbitrary reference temperature T_R. The value of β_T may be determined from the measurement of G_∞ with temperature. The value of α_T must be deduced from the variation of viscosity with temperature, using the following argument. Let η_0 be the viscosity at the reference temperature so that $[\tau_m]_0 = \eta_0/[G_\infty]_0$. Then the value of τ_m at temperature T is given by

$$[\tau_m]_T = \frac{\eta_T}{[G_\infty]_T} = \frac{\eta_T}{\beta_T [G_\infty]_0}$$

But $[G_\infty]_0 = \eta_0/[\tau_m]_0$ so that

$$[\tau_m]_T = \frac{\eta_T}{\eta_0} \frac{[\tau_m]_0}{\beta_T}$$

and so
$$\alpha_T = \frac{[\tau_m]_T}{[\tau_m]_0} = \frac{1}{\beta_T} \frac{\eta_T}{\eta_0} \qquad (3.44)$$

Therefore α_T can be determined from β_T and the ratio of the viscosities at the measurement temperature and the reference temperature. As the latter term varies much more rapidly with temperature, the effect of β_T may sometimes be ignored with little loss in accuracy. Using equations (3.43), equation (3.14) for the storage modulus may be written

$$[G'(\omega)]_T = [G_\infty]_T \frac{\omega^2([\tau_m]_T)^2}{1+\omega^2([\tau_m]_T)^2}$$

$$= \beta_T [G_\infty]_0 \frac{(\alpha_T\omega)^2([\tau_m]_0)^2}{1+(\alpha_T\omega)^2([\tau_m]_0)^2}$$

or
$$\frac{[G'(\omega)]_T}{\beta_T} = [G'(\alpha_T\omega)]_0 \qquad (3.45)$$

That is, G' measured at frequency ω and temperature T is equivalent, when divided by β_T, to G' measured at a frequency $(\alpha_T\omega)$ and the reference temperature T_R. Data obtained at difference temperatures may be plotted on a single curve, representing the behaviour at the reference temperature, by using the reduced variables $G'(\omega)/\beta_T$ and $\alpha_T\omega$. Similar expressions may be derived for all the other viscoelastic functions; these are given in Table 3.2.

The analysis above has been carried out using a single Maxwell element as a model for the liquid. The behaviour of an actual material, represented by a series of Maxwell elements, or in a limit a continuous relaxation spectrum, may be treated in the same way if the following two conditions are satisfied. First, the amplitudes of all the components of the spectrum H must have the same temperature dependence, i.e., they must all be described by the same β_T. Second, the relaxation time of all the components of the spectrum must have the same temperature dependence, i.e., they must all be described by the same α_T. These two conditions are equivalent to saying that the shape and width of $H(\tau)$ must not change with temperature, although the overall amplitude may change and the position of $H(\tau)$ on the

Table 3.2. Reduced variables, denoted by subscript p, which enable data obtained at different temperatures to be reduced to a single curve:

$\beta_T = [G_\infty]_T / [G_\infty]_0 \, ; \, \alpha_T = \eta_T / \eta_0 \beta_T$.

Subscript T refers to measurement temperatures, subscript 0 to reference temperature.

$G_p'(\omega) = G'(\omega)/\beta_T$	vs	$\omega \alpha_T$
$G_p''(\omega) = G''(\omega)/\beta_T$	vs	$\omega \alpha_T$
$\eta_p'(\omega) = \eta'(\omega)/\alpha_T \beta_T$	vs	$\omega \alpha_T$
$G_p(t) = G'(t)/\beta_T$	vs	t/α_T
$J_p'(\omega) = J'(\omega)\beta_T$	vs	$\omega \alpha_T$
$J_p''(\omega) = J''(\omega)\beta_T$	vs	$\omega \alpha_T$
$J_p(t) = J(t)\beta_T$	vs	t/α_T

time scale may change. It follows that the reduced parameters for the relaxation spectrum are $H(\tau)/\beta_T$ and τ/α_T. Similar arguments and conditions apply to the generalised Voigt model, and the reduced variables for the retardation spectrum are $L(\tau)\beta_T$ and τ/α_T. The factors α_T and β_T may thus be placed outside the integrals of equations (3.26) to (3.39) and the expressions given in Table 3.2 used for these generalised models also.

An alternative method of reducing data is to use normalised variables, each function being divided by the limiting value at the reference temperature. The storage modulus is then plotted as $G'(\omega)/G_\infty$ against the parameter $\omega \tau_m = \omega \eta / G_\infty$, all values being taken at the temperature of measurement. The other functions are normalised in a similar manner.

The use of pressure as an additional variable to temperature and frequency may also be treated in the same way, using α and β factors to account for the variation of relaxation time and modulus with pressure. Normalised variables may also be used in this case, the data being normalised with respect to the limiting value of the function of the temperature and pressure of measurement. The conditions regarding changes of the shape and width of the spectrum with temperature as discussed previously apply equally to changes arising from variations in pressure if the data are to be satisfactorily reduced to a single curve.

4. High Frequency Methods for Measuring the Mechanical Properties of Liquids

In order to observe viscoelastic effects in a material, the time scale of any applied deformation must be comparable with the time associated with molecular rearrangements. For supercooled liquids, the Maxwell relaxation time $\tau_m = \eta/G_\infty$ (equation 3.12) provides a convenient measure of this time, in spite of the inadequacies of the Maxwell element as a model for the viscoelastic behavior of a liquid. For a typical value for the limiting high frequency shear modulus, G_∞, of 10^9 Pa, the relaxation time of a supercooled liquid is $\tau_m = \eta \times 10^9$ s, where η is the viscosity in Pa s. In stress relaxation and creep measurements, and low frequency alternating stress measurements, the experimental time scale is in the range 10^{-2} to 10^3 s. These techniques therefore necessitate measurements being made on liquids having viscosities in the range 10^7 to 10^{12} Pa s if viscoelastic effects are to be observed. Consequently, the measurements must be made at, or near, the glass transition temperature of the liquid. The use of alternating stress measurements at higher frequencies allows measurements to be carried out at lower values of viscosity and thus at higher temperatures: this is often experimentally more convenient.

Two further considerations indicate the desirability of operating at high frequencies. Firstly, the method of reduced variables allows, in principle, the use of any frequency of measurement by making the measurements at an appropriate temperature. However, it is necessary to demonstrate the validity of the assumptions inherent in the method of reduced variables by obtaining measurements over as wide a frequency range as possible, as well as over a range of temperatures. Secondly, in order to determine reliable values for the limiting high frequency shear modulus, and its temperature variation, it is necessary to use as high a frequency of measurement as is possible. For these reasons, frequencies in the megahertz range are widely used in studying the viscoelastic behavior of supercooled liquids, and this chapter is largely devoted to the description and analysis of techniques operating in this frequency range. Techniques operating at frequencies below 1 MHz have found little application in the study of supercooled liquids, although they are widely used for polymers and polymer solutions. They have been used, however, to investigate the effects of polymer additives to lubricating oils, both as a function of temperature and pressure (Philippoff, 1963).

The generation of mechanical vibrations and waves at these frequencies is possible only by using magnetostrictive or piezo-electric materials. Present day techniques are largely derived from the pioneering work of W. P. Mason *et al.* (1949) and McSkimin (1952) at the Bell Telephone Laboratories, in the use of piezo-electric transducers for the measurement of liquid properties. At the present time, the maximum frequency at which reliable measurements can be made is of the order of 1000 MHz (Lamb and Richter, 1966). At this frequency, viscoelastic relaxation can be studied in liquids having viscosities of the order of 1 Pa s or greater. In low molecular weight materials, this value of viscosity can only be obtained in liquids which will supercool. The study of the viscoelastic behavior of liquids which crystallize, with a maximum viscosity of perhaps only 0.01 Pa s, awaits the development of still higher frequency techniques.

It has been shown, in Section 3.1, that a shear wave propagated into a viscous liquid is highly attenuated, and the amplitude is reduced to a negligible value in a very short distance. Measurements of the velocity and attenuation of the wave in the liquid are therefore not usually feasible. Instead the shear wave may be propagated in a low loss solid material, either as a guided wave or as a free-space wave. Interaction between the wave and the liquid then occurs at an interface between the liquid and the material in which the wave is propagated. In the case of a guided wave, propagated along a wave guide in the form of a bar or strip, the change in the propagation constant produced by loading the surface of the wave guide with the liquid is measured. For a free-space wave, the reflection coefficient at a solid-liquid interface is measured. Techniques involving the measurement of the reflection coefficient are described in Section 4.1, and techniques involving guided waves are described in Section 4.2. A third type of technique, using a transducer continuously driven at its resonant frequency, is described in Section 4.3. The change in the resonant frequency and the effective change in the mechanical losses are measured when the transducer is immersed in the liquid.

The attenuation of longitudinal plane waves propagated in a liquid is much less than that for shear waves. It is therefore possible to determine the attenuation and propagation velocity of the wave directly by using a variable path pulse technique. This technique is described in Section 4.4. The relevant theory and the method of calculating the complex bulk modulus from the measurements is given in Chapter 6.

Techniques using the interaction between elastic waves, both longitudinal and shear, and optical waves are described in Section 4.5. These techniques offer the possibility of obtaining measurements at frequencies up to 10 GHz, and thus complement the conventional techniques operating at lower frequencies.

All the techniques described in this chapter operate at frequencies above 10^4 Hz. Many techniques operating at lower frequencies have been developed, but they are more suited to measurements on viscous materials such as high molecular-weight polymers, and rubbery materials, than on supercooled liquids. An excellent survey of such techniques is given by Ferry (1970, Chap. 5).

4.1 Reflection Coefficient Techniques

When a plane shear wave is reflected at an interface, the boundary conditions require, in general, that both longitudinal and shear reflected and transmitted waves are produced. The exception to this general condition is a shear wave with the particle motion parallel to the surface (Thurston 1964 p. 79). Then both the reflected and transmitted waves are plane shear waves, the angle of incidence is equal to the angle of reflection, and the angle of refraction is given by Snell's Law. The simplest case, that of normal incidence, when the reflected and transmitted waves are also normal to the surface is analysed in the following section (Kinsler and Frey, 1962 p. 128).

4.1.1 REFLECTION OF A PLANE SHEAR WAVE AT AN INTERFACE, FOR NORMAL INCIDENCE

Consider a shear wave, travelling in the positive y direction in medium 1, and incident normally on the interface with medium 2 (Fig. 4.1). The two media are each characterised by a characteristic shear impedance $Z = \sqrt{(\rho G^*(j\omega))}$, where ρ is the density, and $G^*(j\omega)$ is the complex, frequency dependent, shear modulus. It follows that the impedance Z is in general a complex quantity. The incident wave may be specified by the shear stress σ_i, given by the equation similar to equation (3.3).

$$\sigma_i = \sigma_1 \exp[j(\omega t - \Gamma_1 y)] \quad (4.1)$$

where $\omega = 2\pi f$ is the angular frequency and Γ_1 is the propagation constant, given by $\Gamma_1 = \omega/c_1$ where $c_1 = \sqrt{(G_1^*(j\omega)/\rho_1)}$. At the interface, part of the energy is transmitted into medium 2 as a transmitted wave,

$$\sigma_t = \sigma_3 \exp[j(\omega t - \Gamma_2 y)] \quad (4.2)$$

where Γ_2 is the propagation constant for medium 2.

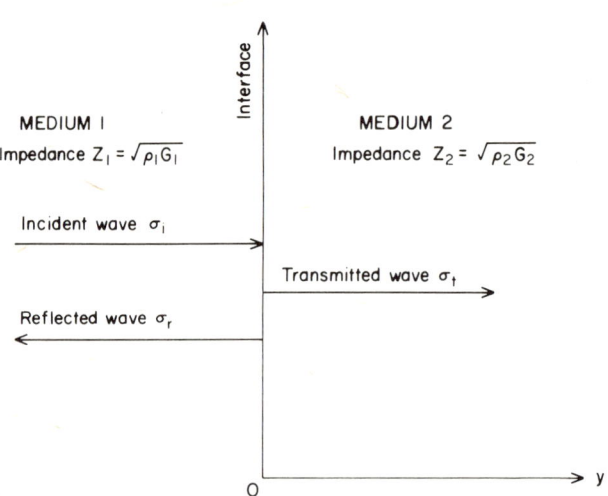

FIG. 4.1. Reflection and transmission of waves at impedance discontinuity.

4. HF METHODS FOR MEASURING MECHANICAL PROPERTIES

The remainder of the energy is reflected back into medium 1 as a reflected wave, travelling in the negative y direction, given by

$$\sigma_r = \sigma_2 \exp[j(\omega t + |\Gamma_1 y)] \qquad (4.3)$$

The amplitude terms σ_1, σ_2 and σ_3 all refer to the peak amplitudes of the stress waves at the arbitrary co-ordinate origin $y = 0$.

At the interface, the following two conditions must be satisfied: (1) continuity of particle velocity (or displacement) and (2) continuity of stress.

Condition (2) is satisfied by summing and equating the stresses in the two media at the interface, which may be put at the point $y = 0$. Then $\sigma_i + \sigma_r = \sigma_t$, evaluated at $y = 0$, which gives

$$\sigma_1 + \sigma_2 = \sigma_3 \qquad (4.4)$$

To satisfy condition (1) it is necessary to specify the velocities in terms of the shear stresses. For a forward travelling wave, the displacement ξ is given by equation (3.3)

$$\xi = \xi_0 \exp[j(\omega t - |\Gamma y)]$$

and the particle velocity is $\dot\xi = j\omega \xi$. The strain $\gamma = \partial \xi / \partial y$ is related to the stress by the shear modulus $G^*(j\omega)$ so that

$$\sigma = G^*(j\omega) \frac{\partial \xi}{\partial y}$$

or
$$\sigma = G^*(j\omega)(-j\Gamma)\xi = -\frac{|\Gamma|}{\omega} \dot\xi G^*(j\omega)$$

But $|\Gamma/\omega = \sqrt{(\rho/G^*(j\omega))}$, so that the stress and particle velocity are related by the equation $\sigma = -\sqrt{(\rho G^*(j\omega))}\dot\xi$. The quantity $\sqrt{(\rho G^*(j\omega))}$ is the characteristic shear impedance Z, so that for a forward travelling wave, the particle velocity is given by

$$\dot\xi = -\frac{\sigma}{Z}$$

Thus for the incident and transmitted waves,

$$\dot\xi_i = -\frac{\sigma_i}{Z_1}$$

and

$$\dot{\xi}_t = -\frac{\sigma_t}{Z_2}$$

For a backward travelling wave, the displacement is given by

$$\xi = \xi_0 \exp[j(\omega t + \Gamma y)]$$

so that the stress is given by

$$\sigma = G^*(j\omega)\frac{\partial \xi}{\partial y} = G^*(j\omega)(+j\Gamma)\xi = +\sqrt{(\rho G^*(j\omega))}\dot{\xi}$$

Thus for the backward travelling reflected wave in medium 1 the particle velocity is given by

$$\dot{\xi}_r = +\frac{\sigma_r}{Z_1}$$

Boundary condition (1) is satisfied by equating the particle velocities in the two media at the interface

$$\dot{\xi}_i + \dot{\xi}_r = \dot{\xi}_t$$

or in terms of the stresses,

$$-\frac{\sigma_1}{Z_1} + \frac{\sigma_2}{Z_1} = -\frac{\sigma_3}{Z_2} \qquad (4.5)$$

Combining equations (4.4) and (4.5) and eliminating σ_3 gives

$$\frac{\sigma_1 - \sigma_2}{Z_1} = \frac{\sigma_1 + \sigma_2}{Z_2}$$

or after rearranging,

$$\frac{\sigma_2}{\sigma_1} = \frac{Z_2 - Z_1}{Z_2 + Z_1} \qquad (4.6)$$

Thus the stress amplitude reflection coefficient at the interface, σ_2/σ_1 is related to the impedances of the two media. If Z_1, for the material in which the wave is propagated, is known then Z_2, the impedance of medium 2, can be determined from a measurement of the reflection coefficient. If medium 1 a solid in which the wave propagates with very little loss, such as fused quartz, then Z_1 is a real

4. HF METHODS FOR MEASURING MECHANICAL PROPERTIES

quantity. If medium 2 is a liquid, then Z_2 will be complex and small compared with Z_1, so that the reflection coefficient, also a complex quantity, has a value close to -1. It is convenient to express the reflection coefficient as

$$\sigma_2/\sigma_1 = R\underline{/\pi - \theta} = -R\cos\theta + jR\sin\theta$$

where the magnitude R is just less than unity, and θ is a small positive angle. Equation (4.6) may be rearranged to give an expression for Z_2,

$$Z_2 = Z_1 \frac{1 + \sigma_2/\sigma_1}{1 - \sigma_2/\sigma_1} = Z_1 \frac{1 - R^2 + j2R\sin\theta}{1 + R^2 + 2R\cos\theta} \quad (4.7)$$

The plane wave shear impedance of a liquid has been defined as $Z_L = R_L + jX_L$. If θ is small, so that $\cos\theta \simeq 1$, then equation (4.7) becomes

$$Z_L = R_L + jX_L = Z_1 \left\{ \frac{1 - R^2}{(1 + R)^2} + j\frac{2R\sin\theta}{(1 + R)^2} \right\} \quad (4.8)$$

The error introduced by the assumption that $\cos\theta = 1$ is less than 1% for most liquids, as in most cases $Z_L < 0.1\, Z_1$, and equation (4.8) may be used to evaluate the components of Z_L. The use of this equation also enables R_L to be calculated from a knowledge of the magnitude of the reflection coefficient only, as

$$R_L = Z_1 \left\{ \frac{1 - R^2}{(1 + R)^2} \right\} = Z_1 \left\{ \frac{1 - R}{1 + R} \right\}$$

The components of the shear modulus $G^*(j\omega) = G'(\omega) + jG''(\omega)$ can be calculated from R_L and X_L using equations (3.7) and (3.8);

$$G'(\omega) = \frac{R_L^2 - X_L^2}{\rho} \quad (3.7)$$

$$G''(\omega) = \frac{2R_L X_L}{\rho}. \quad (3.8)$$

4.1.2 NORMAL INCIDENCE PULSE TECHNIQUE

The analysis of the reflection of a plane wave at an impedance discontinuity given in the previous section assumed steady-state conditions in semi-infinite media, in that further reflections of the waves at the boundaries of the media were ignored. This condition is fairly readily realised if medium 2 is a liquid, as any waves transmitted into the liquid are absorbed in a very short distance. Medium 1 is normally a material, of a finite size, in which a shear wave can propagate with very little loss. The reflected wave σ_r in Fig. 4.1 will therefore travel to the generating surface where, upon reflection, a second incident wave is formed. When this wave reaches the interface, further transmitted and reflected waves are generated, and so the process continues, with a decrease in amplitude of the wave at each reflection. In the steady state, all the forward travelling waves can be summed and represented as a single incident wave, and a similar argument can be applied to the reflected and transmitted waves. The situation shown in Fig. 4.1 can be used to represent these three waves, and the same analysis used. The combination of the incident and reflected waves in medium 1 sets up a standing wave pattern, with a succession of stationary maxima and minima, each separated by half a wavelength. Measurements of the intensity and position of the standing wave pattern enable the reflection coefficient at the interface to be determined (Kinsler and Frey, 1962 p. 134). This method is widely used in liquids and gases with longitudinal waves, where the wave pattern is readily measured (Hubbard, 1931; Greenspan, 1964). In a solid, the amplitude of a wave can only be measured by attaching a suitable transducer to a reflecting surface and measurement of the standing wave pattern is not possible.

Instead, the amplitudes of the incident and reflected waves are measured using a pulse technique which time separates the waves. A transducer bonded to one end of a solid rod radiates a short train of shear waves into the rod, and then later serves as a receiver for the waves reflected from the far end of the rod. Successive reflections at both ends of the bar result in a received signal consisting of a series of echoes of diminishing amplitude. The process is repeated periodically at a rate sufficiently slow that interference between successive trains of echoes is avoided. The duration of the train, or pulse, of shear waves must be shorter than the time required for the wave to travel down the rod and back if successive reflections are to be separated in time. The propagation velocity of a free space shear wave is given by $c = \sqrt{(G/\rho)}$ where G is the shear modulus and ρ is the density of the material. For fused quartz, $G = 31.2 \times 10^9$ Pa and

$\rho = 2200 \text{ kg m}^{-3}$, giving a velocity of $c = 3.76 \times 10^3 \text{ m s}^{-1}$. A quartz rod 5 cm long would therefore have a transit time of $10/(3.76 \times 10^5) \text{ s} = 26 \,\mu\text{s}$. Thus, a pulse of duration less than 26 μs would be necessary to ensure separation of the echoes. The pulse must be long enough for steady state conditions to be established after the decay of the initial transients. Pulse lengths in the range 5 to 10 μs are typically used. A 10 μs duration pulse at a frequency of 30 MHz contains 300 cycles, so that steady state conditions exist for the major portion of the pulse. Figure 4.2 shows schematically the received signal from a 5 cm delay rod, for a pulse width of 10 μs; the envelope only of each reflected echo is shown. The exponential decay in amplitude of the pulses is due to several sources of energy loss, including losses in the fused quartz bar, losses in the bond between the transducer and the bar, and energy absorbed by the transducer and the receiving system. This small "background" attenuation is present when no liquid is present at the end of the rod. Under these conditions, Z_2, for air, is negligible so that $\sigma_2/\sigma_1 = -1$ or $R = 1$, $\theta = 0$. If now liquid is applied to the end of the rod, R and θ will change in value R becoming just less than unity, and the attenuation of the pulse train will be increased. The change in the attenuation per reflection is used to determine the magnitude of the reflection coefficient, R, from which the real part of the shear impedance R_L may be calculated. The increase in the value of θ results in a change in the phase of the train of oscillations within each pulse relative to the transmitter pulse. Measurement of this

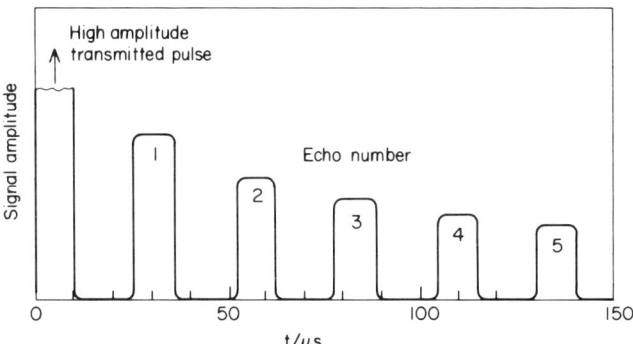

FIG. 4.2. Schematic diagram of received signal from a 5 cm fused quartz bar, with a pulse duration of 10 μs.

phase change enables the imaginary part of the shear impedance, X_L, to be calculated. The phase change per reflection is normally small, as the value of X_L is always very much smaller than the impedance of the fused quartz: the maximum value of X_L in the center of the relaxation region is typically less than 2×10^5 N s m^{-3} (Barlow et al. 1967a), the value of Z for fused quartz is 8.3×10^6 N s m^{-3} ($= \sqrt{\rho G}$). The phase shift is consequently of the order of 2° or less, and measurements of acceptable accuracy are obtainable only with difficulty. Techniques using normal incidence of the shear waves at the solid-liquid interface are therefore usually restricted to the measurement of R_L only. This restriction is acceptable in the region where X_L is less than about a fifth of R_L and its contribution to $G'(\omega)$ is small; equation (3.7) then reduces to

$$G'(\omega) = R_L^2/\rho$$

A modified system in which the incident waves impinge obliquely on the interface offers increased sensitivity and enables satisfactory measurements to be made of both R_L and X_L.

4.1.3 INCLINED INCIDENCE TECHNIQUE

O'Neill (1949) has shown that for the case of a shear wave incident on an interface with an angle of incidence ϕ, the shear impedance of the liquid is related to the reflection coefficient by the equation

$$Z_L = R_L + jX_L = \frac{\cos\phi}{\cos\psi} \frac{1 - R^2 + j2R \sin\theta}{1 + R^2 + 2R \cos\theta} \qquad (4.9)$$

where ψ is the angle of the refracted wave in the liquid to the normal. To maintain pure shear conditions, the particle motion must remain parallel to the surface and normal to the direction of propagation. For the usual range of values of Z_L, i.e. $Z_L < 0.1 Z_1$, and $X_L \leqslant R_L$, the angle ψ is small, and $\cos\psi$ can be taken to be unity with negligible loss in accuracy. Then for a value of ϕ of around 78°, $\cos\phi \simeq 0.2$, and the system is five times more sensitive than the normal incidence system. Useful measurements of the phase change θ are possible, using the system shown in Fig. 4.3. Two fused quartz bars are used, of the same dimensions, each with a shear transducer bonded to one end. The transducers are excited by a short pulse of high frequency oscillation, and a beam of shear waves is

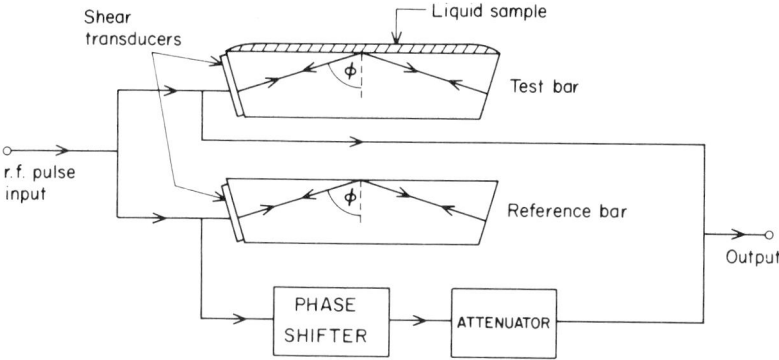

FIG. 4.3. Principle of phase measurement system, inclined incidence technique.

generated in the bar. After reflection at the upper surface and the end of the bar, the pulse of shear waves returns to the transducer, being reflected a second time from the upper surface, and generates an electrical signal. Successive reflections give a train of pulses of decreasing amplitude. The signal received from the second bar is passed through a variable attenuator and a phase shifter and is used as a phase reference signal for the signal from the test bar. The two signals are added together in the receiving system, and the attenuator and phase shifter adjusted so that for a given echo the signals from the two bars are equal in amplitude and opposite in phase. The two pulses therefore cancel, apart from the transients present at the beginning and end of the pulses. The liquid under test is now applied to the upper surface of the test bar. The phase change of the signal from the test bar is then determined from the change in the setting of the phase shifter needed to maintain cancellation of the two pulses.

4.1.4   ELECTRICAL SYSTEM

Figure 4.4 shows the system used for the measurement of the changes of amplitude and phase of the electrical output signals which result when a liquid is applied to the reflecting surface of a fused quartz bar. Components a to g are common to both normal and inclined incidence systems. Components h and i and the reference acoustic system are used for phase measurement in the inclined incidence system only.

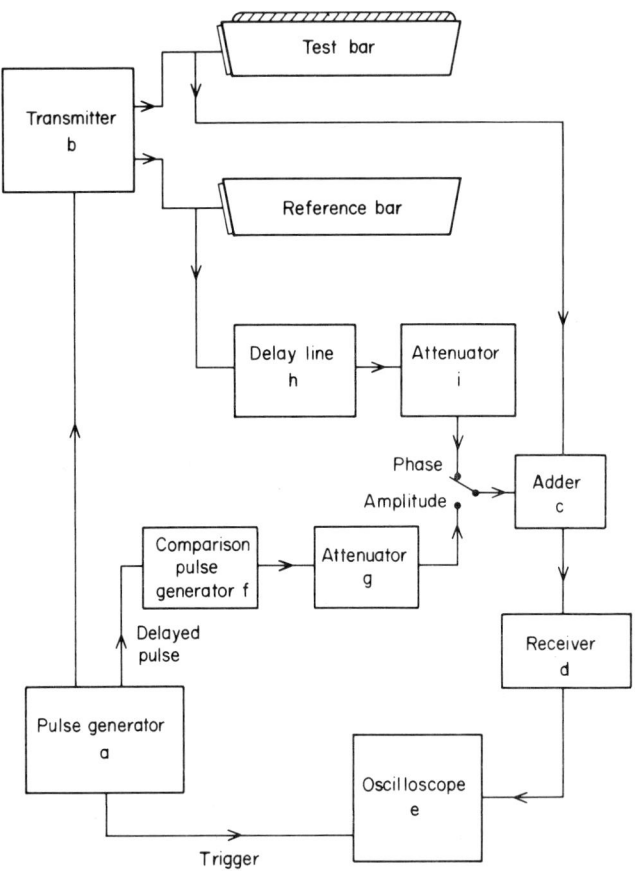

FIG. 4.4. Schematic diagram of the electrical system for inclined incidence system (after Barlow and Lamb, 1959).

The pulse generator a provides trigger pulses for the transmitter b, the oscilloscope e and a delayed trigger for the comparison pulse generator f. The repetition rate is chosen to be sufficiently low that all the reflected pulses resulting from one transmitter pulse are completely attenuated before the transducer is excited again. Operation in the range 50 to 200 Hz usually satisfies this requirement and is rapid enough to avoid excessive flicker when viewing the oscilloscope display. When triggered by the pulse generator, the transmitter b generates a short pulse of high frequency operation, of between 5 and 20 μs duration and a peak amplitude of between 30 and 100 volts. This pulse is applied directly to the transducer, a BT or BC cut quartz crystal plate (Cady, 1946) which vibrates in the thickness shear mode, and may be used at the fundamental resonance or at an odd harmonic of the fundamental, the transmitter frequency being chosen accordingly. The series of received pulses resulting from the multiple reflections in the bar is amplified by the receiver d and after detection is displayed on the wide band oscilloscope e. A pulsed reference signal, at the same frequency as the transmitted signal, is generated by the oscillator f, passed through the attenuator g and added to the received signal in the adding unit c. By adjustment of the delayed trigger pulse, the reference pulse may be displayed adjacent to any of the received pulses, and may be made equal in height by adjustment of the attenuator setting. The direct connection between transmitter output and the receiving system, required by the use of a single transducer to both transmit and receive the pulse of shear waves, places special requirements on both the transmitter and receiver. The adding unit and receiver must be able to recover rapidly from the high amplitude transmitter pulse, so that the reflected echoes can be received without distortion. A recovery time for the system of about 1 μs is possible with suitable design. Any spurious output from the transmitter must be negligible compared with the received signal from the transducer. This generally requires that the output during the "off" period is more than 160 dB, or a factor of 10^8, less than the output pulse amplitude. Circuit details of transmitters having this high rejection ratio are given by McSkimin (1961, 1962), Barlow and Subramanian (1966).

The procedure for measuring R_L consists of noting the setting of attenuator g for which the comparison pulse is equal in height to a received pulse. After applying liquid to the surface of the bar (the end of the bar in the normal incidence system) the process is repeated and the difference in the attenuator setting is noted. Measurements are typically made on the first four to six received pulses, and the average amplitude change per reflection determined by graphical or other means. R_L is then calculated using equation (4.8).

The phase shifter h, used in determining the phase change of the received pulses, consists of lengths of precision co-axial cable arranged in two series of lengths, differing by 100 cm and 10 cm. A constant impedance adjustable line is used to obtain a fine adjustment of the total delay. The phase change resulting from a given length of cable is determined by the propagation constant of the cable. The delay line and attenuator i are adjusted to obtain the cancellation of a received pulse, the amplitude comparison pulse being switched out of circuit during this operation. The liquid is then introduced onto the test bar, and the change in delay line setting needed to restore cancellation is used to determine the angle θ and hence the value of X_L. Measurements are again made on the first four to six pulses and an average value of the phase shift per reflection is computed. It is possible to use the change in the setting of attenuator i in this process to determine the amplitude change of the pulses. The method is not used in practice because of small amplitude changes produced by changes in the delay line length: the use of a separate comparison pulse is found to give more reliable attenuation readings. It is essential that no change in the phase of the reference signal occurs when the setting of attenuator i is altered. For this reason, a piston attenuator is used. This enables the attenuation to be continuously changed, with a resolution of ±0.03 dB, with negligible phase change. The principles of operation and design details are given by Moreno (1958), Andreae et al. (1958).

The inclined incidence measuring system described in this section has been found capable of measuring the real part of the shear impedance, R_L to an accuracy of better than $\pm 3 \times 10^3$ N s m^{-3}, and the imaginary part, X_L to an accuracy of better than $\pm 5 \times 10^3$ N s m^{-3} up to a frequency of 35 MHz, and to about $\pm 10^4$ N s m^{-3} at 75 MHz. With the normal incidence system, R_L can be determined to an accuracy of better than $\pm 1.5 \times 10^4$ N s m^{-3}. The frequency range of the technique, from 5 MHz to 75 MHz, is determined largely by the acoustic system, as discussed in the following section.

4.1.5   ACOUSTIC SYSTEM

Fused quartz is commonly used as the material in which the shear waves are propagated. It is resistant to chemical attack from most materials, has good dimensional stability and is easily worked to the optical tolerances on the dimensions that are required. Also, the attenuation of shear waves is comparatively low at frequencies below 100 MHz. The length of a delay rod is a compromise between the requirements of adequate pulse duration and echo separation, and increasing attenuation at the higher frequencies. Lengths in the range 5 cm to 8 cm are found to be convenient. The cross-sectional dimensions are determined by the divergence of the beam of shear waves (Kinsler and Frey, 1962 p. 176), this effect being most severe at the lowest frequency of operation. The radiation pattern in front of a flat source is essentially a parallel beam, of the same diameter of the source, for a distance of approximately a^2/λ, where a is the radius of the source and λ is the wavelength. In this region, the Fresnel zone, the wave propagates with a plane wave front. Further from the source, in the Fraunhoffer zone, the radiation begins to spread out, and propagates as a spherical wave in a cone of semi-angle β where $\sin\beta = 1.22\lambda/2a$. If the delay rod has a diameter close to that of the transducer, at distances greater than a^2/λ from the source reflections at the side walls occur, resulting in the generation of spurious signals and an increased attenuation of the signal. There is thus a lower limit to the diameter of the bar if beam spreading effects are to be held to an acceptable level, a width for the transducer of not less than 20λ often being used as a criterion. For a minimum operating frequency of 5 MHz, when the wavelength in the fused quartz is 0.75 mm, the minimum width of the transducer would be 1.5 cm.

The reflecting surfaces must be plane when compared to the wavelength, if the wave is to be reflected without distortion and the wavefront remain plane. The wavelength at 75 MHz is 0.05 mm so that one hundredth of a wavelength corresponds to half a wavelength of visible light, and polishing of the surfaces to optical tolerances is necessary. The angles between the reflecting surfaces must be such that a plane wavefront launched by the transducer into the bar returns after internal reflection as a plane wavefront at the same angle. Otherwise variations in the time of arrival across the wavefront result in partial cancellation of the received signal, giving a reduced output. A difference in angle of 1 minute of arc between the incident and reflected beams results in a path difference of approximately $\lambda/10$ across a transducer 1.5 cm wide at a frequency of 75 MHz. The manufacture of angled quartz bars to angular tolerances appreciably smaller than this is difficult, and it is this fact, together with increased attenuation at high frequencies, which sets an upper frequency limit of about 80 MHz for the system. Some improvement in the angular tolerances can be attained where surfaces are required to be parallel, and the end surfaces of normal incidence bar can be made parallel to within 0.1 minute of arc. Typical specifications for fused quartz bars, based on these considerations, are shown in Fig. 4.5. The non-reflecting surfaces are left rough ground to dissipate any wave striking them and so reduce unwanted reflections.

The transducers are typically BT or BC cut crystal quartz plates with a fundamental resonant frequency in the range 5 MHz to 10 MHz. Using odd harmonics as well as the fundamental gives a frequency range of rather more than a decade, with the frequencies spaced roughly equally on a logarithmic scale. Thus a 6 MHz fundamental crystal has been used at 6, 18, 30 and 78 MHz, and a 5 MHz crystal at 5, 15, 35 and 75 MHz. The bond between the quartz crystal transducer and the fused quartz bar is critical to the satisfactory operation of the acoustic system. The main requirements are that the bond shall be of uniform thickness, thin compared with the acoustic wavelength, and have a high shear modulus. A bond suitable for use over a wide temperature range can be obtained using a cold-welding technique in conjunction with a soft metal having a low melting point, such as indium, or gold alloys (Barlow and Subramanian, 1966). Thin films of metal, about 0.5 μm thick, are applied to one end of the bar and one face of the transducer. The

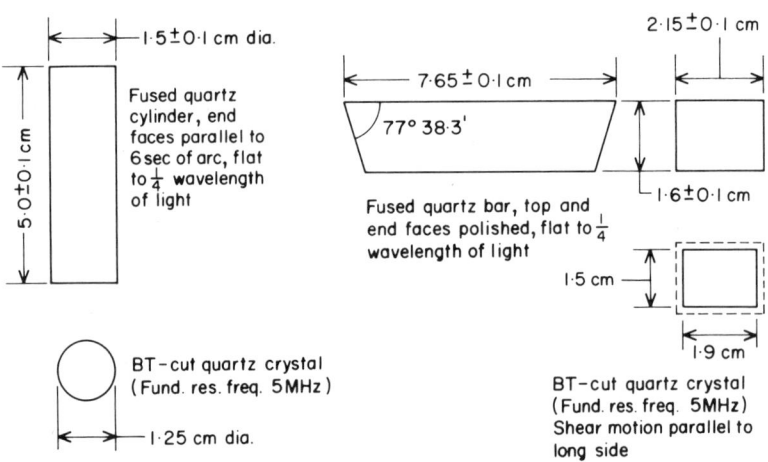

FIG. 4.5. Typical specifications for normal incidence and inclined incidence fused quartz bars (after Barlow and Subramanian, 1966).

two films are then joined by holding them in contact under pressure, at a temperature a few degrees below the melting point of the metal. Great care is required to ensure that no dust particles are present when the two parts are pressed together, as the thin quartz plates are easily cracked. Other materials which have been used are silicone oils, "Nonaq" stopcock grease, paraffin wax, "salol" (phenyl salicilate) and at low temperatures, a variety of supercooled liquids (Matheson, 1971b). These materials enable bonds to be made fairly easily, by pressing and rubbing the transducer on the end of the bar until a very thin film of liquid is obtained. They are less permanent than the metallic bond and may be used over a limited temperature range only. A more permanent bond may be made using an epoxy cement containing a finely ground metal powder.

The transducer is excited by an electric field across the thickness of the plate, produced by applying the pulse of high frequency oscillations to two electrodes, one on each surface of the transducer. The use of a conducting cement, or a metal, for the bonding material enables the bond layer itself to be used as an electrode. Otherwise, a metallic coating of chromium or nichrome evaporated on the end of the bar, prior to bonding the transducer to the bar, is used for one electrode. The second electrode, on the free surface of the transducer, may be formed by a second evaporated metal layer, a conducting paint film, or thin metal foil glued to the surface.

The extent to which temperature stabilisation is required for the acoustic system depends greatly on whether phase measurements are required. The accuracy with which the steady flow viscosity and the shear impedance components are known is such that temperature control of the liquid sample to the order of $0.05°C$ is adequate. However, changes in the velocity of shear waves in fused quartz with temperature changes of this magnitude can cause phase changes larger than those produced by the liquid sample. The first reflected signal in an inclined incidence bar differs in phase from the transmitted pulse by about 3000 wavelengths at 75 MHz, for the dimensions shown in Fig. 4.5. The velocity increases by about 100 p.p.m. per $°C$ temperature rise. Thus, for the resulting phase change to be small compared with that caused by the sample, i.e. to be of the order of $0.3°$ or less, requires a change in the velocity of less than 1 part in 4×10^6. This corresponds to a temperature change of $0.004°C$. Thermal stability of this order may be achieved by placing the acoustic assembly in a container immersed in a constant temperature bath. If the acoustic assembly is insulated from the container by an air space, the small changes in bath temperature are prevented from reaching the bar and liquid. This thermal isolation of the assembly enables the required stability to be attained, but has the disadvantage that several hours are required for thermal equilibrium to be reached. The liquid is held in a small container adjacent to the test bar during this period, so that the liquid is at the same temperature as the bar. A simple mechanical arrangement enables the liquid to be poured onto the test bar after the initial measurements have been completed. The two quartz bars are normally mounted in close proximity, with good thermal contact between them. Some compensation against temperature changes is thus obtained, as both bars respond equally. If only one bar is used, and the reference signal is derived electrically as described by Barlow and Subramanian (1966) the temperature stability requirements are more stringent. The main advantage of using only one quartz bar is that the difficulties of constructing two quartz bars, with transducers matched to operate at the same frequency, are avoided.

4.1.6 VERY HIGH FREQUENCY TECHNIQUES

At frequencies above 100 MHz, the attenuation of the shear waves in the fused quartz becomes appreciable, so that shorter delay bars must be used. Also, the operation of the transducer at high harmonics becomes less efficient, and losses in the bond increase. An alternative method of generating shear waves in a delay rod, which overcomes these difficulties, is to use a rod of piezo-electric material, and to generate the shear wave directly at the surface of the rod (Bommel and Dransfeld, 1960). One end of a suitably oriented quartz crystal rod is inserted into an electromagnetic cavity resonator. The cavity is excited at its resonant frequency by a short burst of high frequency oscillation. The high intensity electric field at the surface of the quartz causes a shear wave to be generated, which propagates along the rod and is reflected each time it reaches an end surface. Each reflection at the sending end re-excites the resonant cavity, giving an electrical output signal which is displayed as an oscilloscope. The end of the quartz bar external to the cavity is coated with liquid, and the increased attenuation of the train of received pulses is measured. The real part of the shear impedance R_L is then calculated using equation 4.8. The orientation of the quartz bar and the angle of the electric field relative to he surface of the bar must be carefully adjusted in order to excite a pure shear mode. These problems are discussed in detail by Lamb and Richter (1966). Using a BC-cut quartz rod, 1 cm in diameter and 1.5 cm in length, as many as 50 echoes may be observed at a frequency of 500 MHz. This technique, using a tunable resonant cavity, has been used over the frequency range 300 to 2000 MHz, with an estimated accuracy in R_L of $\pm 5 \times 10^3$ N s m^{-3}. Measurements of X_L has so far not proved to be possible at these frequencies.

An alternative method of generating shear waves at very high frequencies is to use a thin film transducer of a piezo-electric material. The transducer is evaporated, or sputtered, onto the end surface of the delay rod, and can be oriented to generate either longitudinal or shear waves. The delay rod material need not be piezo-electric but is chosen for its low-loss properties. Shear transducers have been produced on sapphire delay rods using cadmium sulphide, lithium niobate and zinc oxide, and operation at frequencies up to 3 GHz is possible, (Foster *et al.* 1968; Foster, 1969; Duncan *et al.* 1969).

4.1.7 PULSE SUPERPOSITION METHODS

An alternative method of determining the phase angle of the reflection coefficient for a plane wave reflected from a solid-liquid interface has been described by McSkimin and Andreatch (1967a). As in the techniques described in the previous sections, echo decay rates and phase shifts in a delay rod are measured with and without the specimen liquid, from which the complex shear impedance is calculated. Instead of directly measuring the phase change produced by the presence of the liquid, the time delay incurred by an echo making one or more round trips in the delay rod is measured to a high degree of accuracy. To obtain adequate resolution of the small change in the total delay time resulting from the application of liquid to the reflecting surface, a measurement accuracy of the order of 1 part in 10^7 is necessary.

The method consists of generating high frequency pulses of shear waves in the delay rod at a repetition rate f_R such that the time between the pulses is equal to twice the round trip delay time. This method therefore differs from the previously described method in that the train of reflected echoes is not allowed to decay to zero between each transmitter pulse. Consequently interference occurs between the echoes from successive transmitter pulses. The period $T = 1/f_R$ is adjusted until reinforcement of odd numbered echoes is obtained, and the summed echoes have a maximum value. This process results in a high signal-to-noise ratio, with spurious pulses appearing as part of the noise, and the gain of the receiver may be quite low. The waveform resulting from the superposition of the pulses in this way is shown in Fig. 4.6.

The value of the round trip delay T for additive interference between the high frequency waves within each pulse has been given as (McSkimin, 1961)

$$T = p\delta - (p\gamma_0/2\pi f) + n/f \qquad (4.10)$$

where δ is the round trip delay that results from the delay rod alone and p is the number of round trips between superposed echoes; $p = 2$

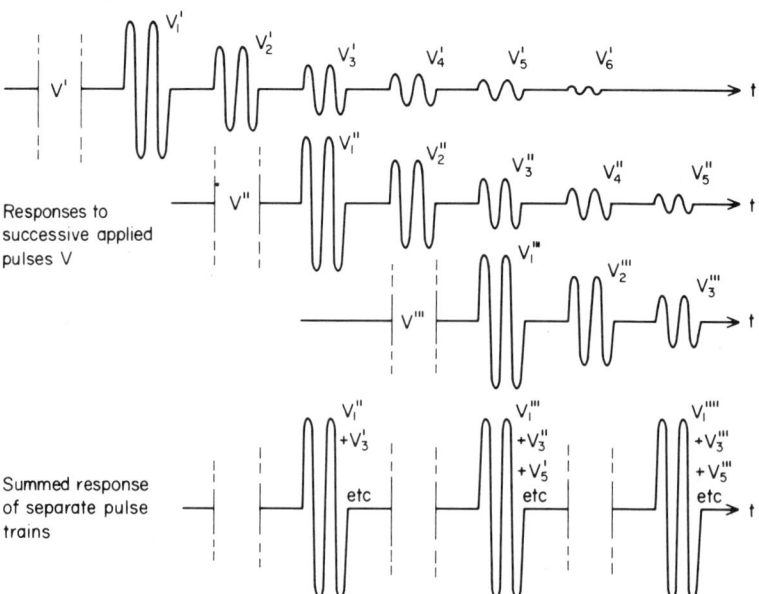

FIG. 4.6. Schematic representation of waves in the pulse superposition method (after McSkimin, 1961).

in this case. γ_0 is a phase angle associated with reflections at both ends of the delay rod, f is the wave frequency and n is an integer equal to the difference in the cycle number in the superposed pulses. A value of T corresponding to $n = 0$ should be used; the difficulties associated with determining the correct value of T are discussed by McSkimin (1961) and Papadakis (1967). When the liquid is applied to the rod, the change θ in the angle of the reflection coefficient changes the angle γ_0 to $\gamma_0 - \theta$. A different repetition rate is thus necessary in order to maintain correct superposition of the pulses. If the delay times corresponding to angles γ_0 and $\gamma_0 - \theta$ are T_0 and T, then $T - T_0 = p\theta/2\pi f$. In terms of the repetition rates $f_{R0} = 1/T_0$ and $f_R = 1/T$, then

$$\theta = \frac{2\pi f}{p} \frac{(f_{R0} - f_R)}{f_{R0} f_R} \qquad (4.11)$$

A frequency synthesiser is used for the repetition rate oscillator, in order to obtain the necessary frequency stability. A highly stable acoustic system is also required, a suitably oriented AT-cut crystal quartz bar being used. This cut of crystal quartz has a lower variation of shear wave velocity with temperature than fused quartz. The system can be automated to automatically adjust the repetition rate to give the maximum amplitude of the summed echoes (McSkimin, 1965; McSkimin and Andreatch, 1967b). Using this technique the shear impedance of liquids as low in viscosity as 0.001 Pa s has been measured at a frequency of 40 MHz, with a resolution in the time delay measurements of 1 part in 10^7. This corresponds to an accuracy in the shear impedance of the order of ± 500 N m s^{-3}.

In a variation of this technique (McSkimin 1970), a sequence of two transmitted pulses is used, the time interval between them being chosen so that a selected echo from the first wave train is superposed on the first echo of the second train. In the time interval between the two transmitter pulses the frequency of the oscillator from which the pulses are obtained is changed by a small amount, and then restored to the original value. This results in a difference in the phase of the wave between the two transmitted pulses. By adjustment of this phase change, the chosen echoes can be placed in phase opposition so that cancellation occurs. The change in phase necessary to maintain the out of phase condition when liquid is applied to the delay rod is then measured. In a system of this type, in which the phase control system was automated, a resolution of 1° in phase change has been achieved, at frequencies up to 500 MHz.

An alternative approach is described by Papadakis (1967) in which a pair of echoes is compared in phase by displaying them in sequence on the oscilloscope, the horizontal display of which is driven by an oscillator at a frequency f_R. The advantage of the high signal-to-noise ratio obtained by the summation of the acoustic waves in the bar is lost, but the circuitry required is considerably simplified. High stability of the acoustic system and the electronic system is again required if adequate resolution is to be obtained.

Various correlation techniques to improve the signal-to-noise ratio have been used, in which long time averaging over many echo trains is used to reduce the effects of random noise (Tittmann and Bommel, 1967; Simmons and Macedo, 1968; McSkimin and Bateman, 1969; Hemphill, 1969). Improvements in signal-to-noise ratio of some 25 dB (a voltage ratio of x16) have been achieved, making measurements possible in situations of high attenuation and high noise level.

4.2 Travelling Wave Techniques

The lower frequency limit of techniques using free space waves is determined by the increasing effects of beam spreading in a delay rod of practical size. Such techniques are therefore rarely used at frequencies much below 5 MHz. In the range 10 kHz to 1 MHz, the use of guided waves is possible, either torsional waves in a cylindrical rod, or plane shear waves in a thin plate. These techniques are operated in a pulsed mode, and the propagation constant of the travelling wave is measured. The change in the propagation constant produced by immersing the rod, or plate, in a liquid can be related to the shear impedance of the liquid.

4.2.1 TORSIONAL WAVE TECHNIQUES

The use of a quartz crystal, vibrating in torsion, to generate torsional waves in a metal rod was first proposed by McSkimin (1952). The transducer is a cylinder of crystal quartz, with the X-axis parallel to the axis of the cylinder (Mason, 1950, p. 91). At the fundamental resonance, the cylinder is a half-wavelength long, with a node in the center. A frequency range of 20 kHz to 100 kHz can be covered with crystals having lengths in the range 10 cm to 2 cm. The crystal is rigidly attached to the end of a long metal rod, of the same diameter at the crystal. Figure 4.7 shows the measuring system used with this technique. The output of a high stability oscillator b is gated to provide pulses containing several cycles of oscillation, which are applied to the transducer via an amplifier d and reed switch e. The gate c is controlled by the pulse generator a. The pulse of torsional waves generated in the rod travels to the end, is reflected and returns to the transducer. During this time interval the reed switch connects the transducer to the receiving system, and the train of reflected

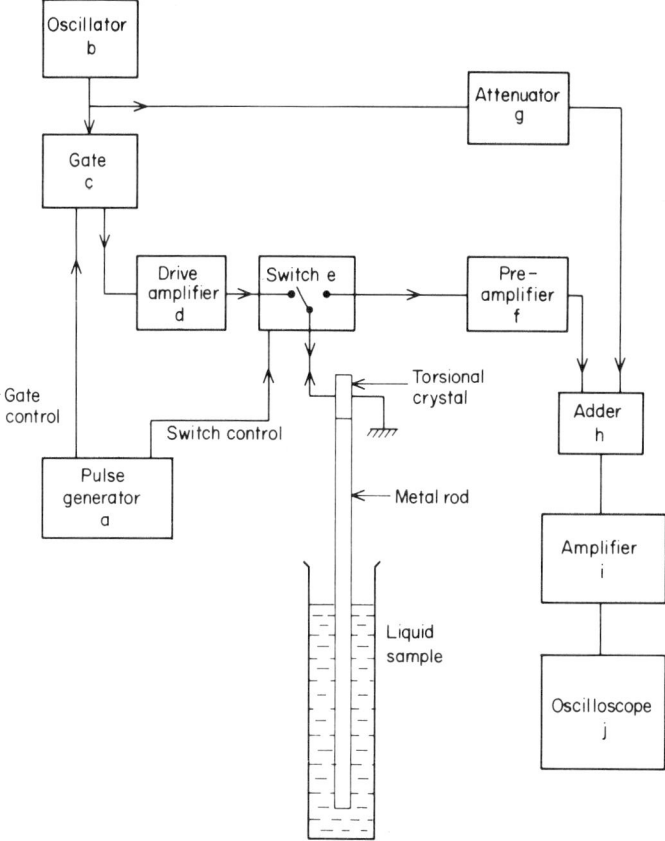

FIG. 4.7. Schematic diagram of measuring system for travelling torsional wave technique.

echoes from successive reflections in the rod is amplified and displaced on the oscilloscope.

For any selected echo, an amplitude and phase reference is established by comparing the received echo against an attenuated continuous wave from the oscillator. By suitable adjustment of the oscillator frequency and the attenuator setting, the two signals can be made equal in amplitude and opposite in phase, giving cancellation for the duration of the pulse. Liquid is now introduced around the rod, and the changes in attenuator setting and oscillator frequency necessary to restore cancellation are determined. The shear impedance of the liquid is given by McSkimin (1952).

$$Z_c = \frac{\rho c r}{2\ell} (\Delta A + j \Delta B) \qquad (4.12)$$

where ρ is the density of the rod material, r is the radius of the rod, c is the velocity of propagation of the torsional wave in the unloaded rod and ℓ is the length of the rod immersed in the liquid. ΔA is the measured change in amplitude per reflection and ΔB is the phase change per reflection, which can be calculated from the measured change in the wave frequency. Z_c is the impedance for the cylindrical torsional waves generated in the liquid, from which the plane wave impedance Z_L is calculated. The two impedances differ significantly only at high values of impedance ($Z_L > 10^4$ N s m^{-3}).

The length of the rod is mainly determined by the need to establish steady conditions during the pulse, and to time separate the successive reflections. The rod material must be resistant to corrosion, and be capable of giving a good surface finish; glass, aluminum, stainless steel and nickel silver have all been used. The velocity of torsional waves in nickel silver, 2×10^3 m s^{-1}, is lower than for many other materials, giving increased sensitivity. A pulse containing 10 cycles at a frequency of 50 kHz would then require a rod at least 40 cm long to ensure time separation of the echoes. The comparatively small number of wavelengths in the rod results in the changes in phase delay caused by the variation of the velocity with temperature being small. The system is not therefore very temperature sensitive, and control to ±0.01°C is found to be adequate.

The approximate analysis used in obtaining equation (4.12) neglects effects due to reflection of torsional waves at the surface of the liquid, and assumes no energy loss on reflection at the end of the rod. These effects can normally be neglected, but are likely to become significant when short lengths of the rod are immersed in liquids of high shear impedance.

Magnetostrictive transducers have been used in a similar technique, using a hollow nickel tube (Glover et al. 1968). Torsional waves are generated in the tube using the Wiedemann effect (Bell et al. 1966). The nickel tube is first magnetised circumferentially by passing along it a short current pulse. The transducer consists of a small coil carrying a pulsed high frequency current. The torsional motion results from the interaction between the magnetic field due to the coil current, and the magnetised rod. The pulse of torsional waves so generated travels along the tube and is reflected from the end. A second coil, placed a short distance below the transmitter coil, is used to detect the return of the echo. A second transmitted pulse of reduced amplitude is used to cancel the returning echo at the receiving transducer (Fig. 4.8). Multiple reflections are prevented by the use of absorbing material placed around the tube above the two transducers. The changes in the amplitude of the second transmitter pulse and the wave frequency, necessary to maintain the cancellation when a portion of the tube is immersed in the liquid to be studied,

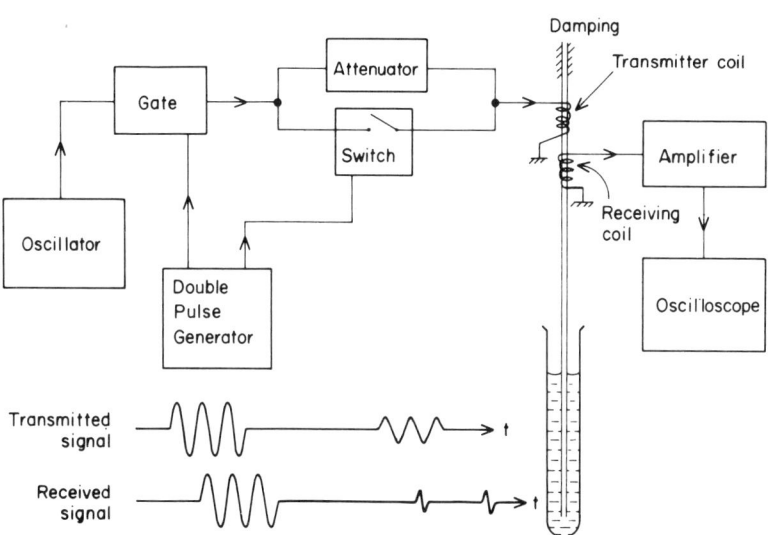

FIG. 4.8. Travelling torsional wave system using magnetostrictive transducers (after Glover et al. 1968).

are used to determine the shear impedance of the liquid. The use of a hollow tube results in an increase in the sensitivity of the system. Equation (4.12), for a solid rod, is modified to give

$$Z = \frac{\rho c r}{2\ell} (\Delta A + j\Delta B) \frac{1 - m^4}{1 + m^3} \qquad (4.13)$$

where m is the ratio of the outer to the inner radii. For a typical value of $m = 0.936$, $(1 - m^4)/(1 + m^3)$ has a value of 0.128 and the sensitivity of the system is increased about eight times. An advantage over the solid rod is that by wetting both sides of the tube, the plane wave impedance is determined directly and so the calculations needed to determine Z_L from the cylindrical impedance Z_c are avoided. Reflection of pulses from the air-liquid interface can again be significant, but the geometry is arranged so that they do not overlap any of the measuring pulses. An accuracy of ± 150 N s m^{-3} is claimed for a system operating in the range 20 to 100 kHz. Continuous frequency coverage is obtained as the transducers are non-resonant. The upper frequency limit is set by the reduced efficiency of the transducers as the axial length becomes comparable to the acoustic wavelength. With transducers 3 mm in length an upper limit of 500 kHz should be possible if a suitable measuring system is used.

4.2.2 TRAVELLING SHEAR WAVE TECHNIQUE

A technique using shear waves in a metal strip delay line has been described by Hunston et al. (1972a, 1972b). A frequency range from 200 kHz to 5 MHz is possible, a range which is not covered by any other technique. A shear wave can be propagated in a thin metal plate, with no dispersion, provided the thickness of the plate is less than a half-wavelength (May, 1964). Under these conditions the displacement is uniform over the thickness of the plate, and is at right angles to the direction of propagation. Such a wave is generated by a shear transducer cemented to the end of a thin metal strip (Fig. 4.9). In aluminum, the wavelength at a frequency of 2 MHz is 1.6 mm necessitating the use of a strip less than 0.8 mm thick. The change in the propagation constant resulting from the immersion of

4. HF METHODS FOR MEASURING MECHANICAL PROPERTIES 89

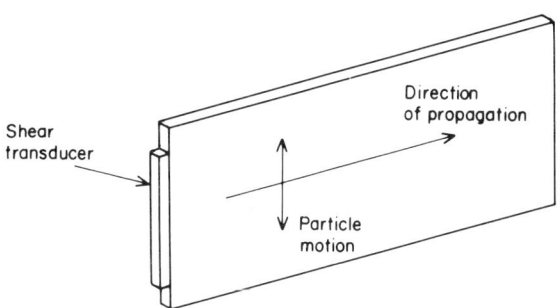

FIG. 4.9. Shear waves in strip delay line.

the strip in a liquid is measured, as in the travelling torsional wave techniques, and the impedance of the liquid is given by an equation similar to equation (4.12),

$$Z_L = \frac{\rho c t}{2\ell}(\Delta A + j\Delta B) \qquad (4.14)$$

where t is the thickness of the strip, and the other quantities are as defined in Section 4.2.1. The amplitude and relative phase of the echoes are measured by the method of Papadakis (1967) described in Section 4.1.7. The repetition rate of the oscilloscope sweep is adjusted to be a multiple of the transit time, so that a pair of echoes is visually superposed on the oscilloscope display. The repetition rate for superposition of a pair of echoes is measured both with and without liquid wetting the delay line, and the phase shift calculated using equation (4.11). A delay time of approximately 50 μs is measurable to an accuracy of ±3 ns. This corresponds to a change in phase angle of ±2°, and an accuracy in X_L of approximately ± 1000 N s m^{-3}. The change in amplitude of the pulses is measured by noting the change in the setting of a series attenuator necessary to maintain an echo at the same amplitude on the oscilloscope display.

A delay line of between 5 and 10 cm long is adequate to obtain time separation of the echoes and to enable steady state conditions to be reached during the pulse. A width of more than 2 cm is necessary to keep beam spreading effects to a reasonable level, and to minimise the appearance of spurious echoes due to the generation of longitudinal waves by reflection of the shear waves at the edges of the strip.

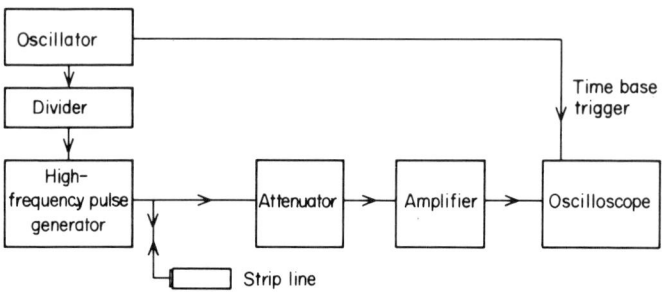

FIG. 4.10. Travelling shear wave in strip delay line. Schematic diagram of measuring system (after Hunston et al. 1972b).

4.3 Resonance Technique

The shear impedance of a liquid can also be determined by measuring the loading effect on a transducer immersed in a liquid. In the frequency range 20 to 100 kHz a quartz crystal vibrating in torsion can be used (Mason, 1947). The cylindrical crystal is used at the fundamental resonance when it is a half-wavelength long. It may be supported at the central nodal region. A length to diameter ratio of 5 to 1 or greater is desirable in order to ensure a fairly pure torsional mode. Measurements are made of the resonant frequency, and the resistance at resonance, first in vacuum and then immersed in the liquid. The changes in the resonant frequency, Δf, and resistance, ΔR, are related to the components of the shear modulus by the equations

$$\Delta R = K_1 R_L$$
$$\Delta f = K_2 X_L \qquad (4.15)$$

The quantities K_1 and K_2 are constants for a given transducer, except at viscosities less than 2×10^{-3} Pa s (Rouse et al. 1950, Barlow et al. 1961). They can be determined either from measurements on liquids of known viscosity and density, or calculated from the dimensions and electrical characteristics of the crystal.

The resistance and resonant frequency are determined using an a.c. bridge (Barlow et al. 1961). Measurements to an accuracy of a few percent are readily obtained on liquids having viscosities less than approximately 1 Pa s. The range of the technique has been extended to 20 Pa s by using a crystal of length to diameter ratio of only 3 to 1 (Philippoff, 1964). The less pure mode of vibration results in some loss of accuracy.

The resonant quartz crystal is readily mounted in a pressure vessel, and then provides a simple means of determining the variation of shear impedance with pressure (Appeldoorn et al. 1962), subject to the same maximum viscosity limitation of about 1 Pa s. The major limitation of this technique is that its use is limited to liquids of low conductivity. The liquid is in contact with the evaporated electrodes of the crystal, and any conduction in the liquid appears as a resistance in parallel with the crystal. This effective resistance reduces the resolution and accuracy of the electrical measurements, especially when the liquid has a high value of shear impedance. It is possible that a thin coating of a hard dielectric material could be evaporated onto the crystal to insulate the electrodes, without unduly affecting the properties of the crystal.

4.4 Longitudinal Wave Techniques

Several different techniques can be used for determining the absorption and velocity of longitudinal waves in liquids, these are reviewed by Matheson (1971a). The pulse technique, first used by Pellam and Galt (1946) and Pinkerton (1947) is the most widely used system for liquids and can be used over the frequency range 0.5 to 300 MHz. A typical arrangement based on the design of Andreae et al. (1958) is shown in Fig. 4.11.

A pulse of longitudinal waves of the desired radio frequency is generated in the lower delay rod by a suitable transducer, such as X-cut quartz. The train of waves, typically of between 10 and 20 μs in duration, passes through the liquid sample and is received by the transducer on the upper delay rod. Only this first pulse is used for measurement purposes, the later signals resulting from reflections within the delay bars being ignored. The amplitude of the first received pulse is measured as a function of the path length in the liquid, thus giving a direct measurement of the attenuation coefficient.

The pulse amplitude is determined by comparing the pulse on the oscilloscope with a pulse of known amplitude. The comparison pulse is adjusted in amplitude by a calibrated piston attenuator, and is

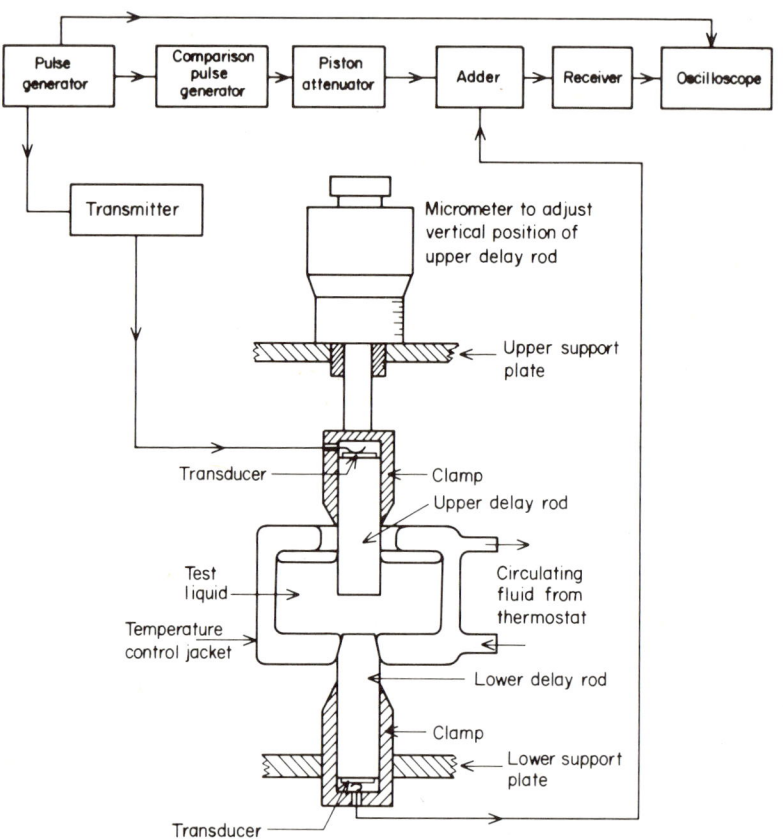

FIG. 4.11. Schematic diagram of variable path-length pulse technique for measuring the longitudinal wave velocity and absorption in liquids.

added to the received signal at the input to the amplifier and detector system. Both pulses, of equal amplitude, thus undergo the same amplification, and errors arising from gain variations and non-linearity in the receiving system are avoided. The output of the piston attenuator is an exponential function of piston displacement. Thus a plot of piston position versus path length in the liquid gives a straight line which has a slope equal to the value of the attenuation coefficient in units of neper (or dB) per cm.

The velocity of propagation of the wave in the liquid can be estimated by allowing successive received pulses, which have made one and three trips through the liquid, to interfere. This is accomplished by increasing the duration of the transmitter pulse so that the trailing edge of the first received pulse overlaps the leading edge of the pulse which has made two further transits of the liquid path. If the liquid path is short compared with the length of the delay rods, this pulse is clearly separated from pulses making multiple reflections in the delay rods. The system can then be used as a variable path interferometer, the beats between the two pulses being counted as the liquid path length is varied by a known amount. The distance between successive maxima then corresponds to one half wavelength. Velocity values accurate to 1 part in 10^3 can be obtained if the attenuation in the liquid is sufficiently low to allow an adequate separation of the delay rods.

The general specifications of the electronic and acoustic systems are similar to those for the shear wave apparatus discussed in Section 4.1.5. The dimensional tolerances specified for the fused quartz bars of the shear wave system apply equally to the longitudinal wave apparatus. The bonding of the transducers to the delay bars is considerably simpler, however, as the high bulk modulus of liquids provides efficient coupling between the transducer and the bar, and a wide variety of materials can be used. Stopcock grease and salol (phenyl salicilate) are perhaps the most useful general purpose materials (Matheson 1971b).

The mechanical alignment of the two fused quartz delay rods is critical. Provision is made for the adjustment of the orientation of the lower delay rod so that the end faces of the delay rods can be made parallel. This condition is indicated by the received signal having the maximum amplitude for a given separation of the rods. The motion of the upper rod must then be constrained so that the end faces remain parallel while the separation is varied. A carriage sliding on ground steel pillars is a convenient method of achieving the required parallel motion. An alternative method, if the moving mass is not too great, is to use parallel cantilever springs as described by Jones (1962).

The spreading of the beam of longitudinal waves due to diffraction requires consideration in a system in which the separation between the source and the receiver is varied. Seki *et al.* (1956) have calculated the variation of amplitude and phase angle across a plane normal to the propagation direction, from which the average pressure over the receiver area can be calculated as a function of distance from the source. If large corrections are to be avoided, the pulse technique should not be used below about 5 MHz unless the delay rods and transducers are made larger than the usual 1—2 cm diameter. Reviews of experimental equipment are given by Andreae et al. (1958); Edmonds et al. (1962); Andreae and Joyce (1962), and Edmonds (1966).

Propagation velocities may be measured to within ±0.2% and absorption coefficients to within ±2% for liquids of low viscosity over the frequency range 5 MHz to 300 MHz. Some deterioration in these accuracy figures occurs with liquids of high viscosity and high absorption.

At the higher frequencies it is more difficult to maintain the required mechanical alignment between the fixed and moving delay rods, and to measure the very small displacements required. Under these conditions a reflection technique can be used to determine the wave velocity. The system is similar to the normal incidence technique described in Section 4.1.2 for shear waves. The change in amplitude of the wave reflected from the end face of a delay rod, when it is immersed in a liquid, is measured. This enables the reflection coefficient and the real part of the acoustic impedance of the liquid to be determined. The complex impedance is given by

$$Z^*(j\omega) = (\rho M^*(j\omega))^{1/2}. \qquad (4.16)$$

The complex propagation velocity is given by equation (6.3) as

$$c_L^*(j\omega) = (M^*(j\omega)/\rho)^{1/2}.$$

so that $\qquad Z^*(j\omega) = \rho c_L^*(j\omega).$

Then from equation (6.5)

$$Z^*(j\omega) = \rho \left[\frac{1}{c_L(\omega)} - j\frac{\alpha}{\omega} \right]^{-1} \qquad (4.17)$$

If it can be assumed that $((\alpha c_L(\omega)/\omega)^2 \ll 1$, then the velocity $c_L(\omega)$ can be determined from the real part of $Z^*(j\omega) = R(\omega) + jX(\omega)$,

$$c_L(\omega) = R(\omega)/\rho \qquad (4.18)$$

Using this technique measurements have been made up to frequencies of 10 GHz (Stewart and Stewart, 1963).

An extremely accurate method of determining the velocity of longitudinal waves in low viscosity liquids has been developed by Barlow and Yazgan (1966). The acoustic system has a fixed path length in the liquid, the spacing between the ends of the two delay rods being set by gauge blocks of known dimensions. A pulse of longitudinal waves at a frequency of 10 MHz is reflected from both quartz/liquid interfaces as shown in Fig. 4.12. The phase difference between the two signals, which corresponds to a double transit of the liquid path, is determined by cancelling each signal in turn against a continuous reference signal which is derived from the same source as

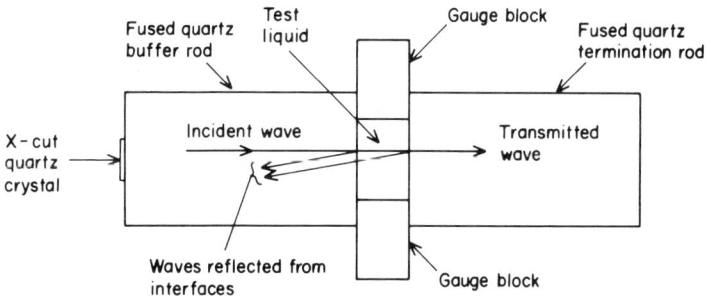

FIG. 4.12. Acoustic system used for the determination of longitudinal wave velocity in liquids by Barlow and Yazgan (1966).

the transmitted signal. The phase of this c.w. signal is altered using a calibrated delay line: the difference between the settings required for cancellation of the two reflected signals, gives the fractional part of a wavelength of the total phase difference. The number of complete wavelengths can be determined by repeating the experiment using a slightly different frequency for which the total number of complete wavelengths is unchanged. The number of complete wavelengths is normally in the range 50 to 300 and can be determined exactly. The fractional part can be determined to $\pm 3°$, giving an overall accuracy in the total phase difference of the order of 3 parts in 10^5 under optimum conditions. Temperature control of the liquid and acoustic system to $\pm 0.001°C$ is necessary in order to obtain measurements of this precision.

4.5 Optical Methods

The interaction between optical and longitudinal waves in liquids was first demonstrated by Debye and Sears (1932) and Lucas and Biquard (1932) who independently showed that light could be diffracted by a beam of longitudinal waves. The waves were generated at frequencies in the megahertz range using a quartz transducer, and were propagated in a variety of low viscosity liquids. Ten years earlier Brillouin (1922) had proposed that in the absence of any external source of longitudinal waves light would be scattered by the random thermal density fluctuations always present in a material. Debye (1912) in his theory of the specific heat of solids, proposed that these thermal motions could be expressed in terms of a series of elastic plane waves. Brillouin predicted that the light scattered by such waves would contain components shifted in frequency by an amount corresponding to the frequency of the elastic wave, the progressive waves acting as a moving diffraction grating. The same result was also predicted by Mandel'shtam in 1918 (Fabelinskii, 1966) and the frequency shifted components are often called the Mandel'shtam–Brillouin components.

When an external source is used to generate a longitudinal wave, at a specific frequency, the technique is referred to as Optical Diffraction. The scattering of light by naturally occurring waves is referred to as Brillouin or Mandel'shtam–Brillouin scattering. In a third technique an intense pulse of light from a laser is focussed within the bulk of a material. Electrostrictive effects produced a longitudinal wave and a reflected light wave is generated which is shifted in frequency by the frequency of the longitudinal wave. This technique is referred to as Stimulated Brillouin Scattering and was first used in crystal quartz and sapphire by Chiao et al. (1964), and in liquids by Brewer and Reickhoff (1964).

Detailed discussion of these techniques and extensive literature reviews are given by Quate et al. (1965), Damon et al. (1970), Smith (1970) and Fabelinskii (1968).

4.5.1 OPTICAL DIFFRACTION

Figure 4.13 shows the experimental arrangement for studying the propagation of longitudinal waves using optical diffraction. A continuous plane longitudinal wave is passed through the liquid

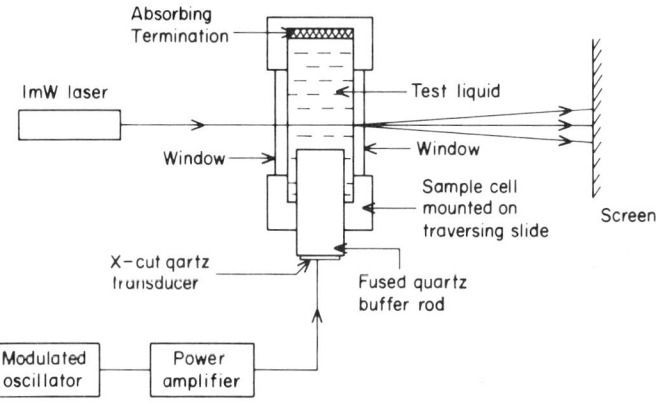

FIG. 4.13. Experimental arrangement for measuring the velocity and absorption of longitudinal waves using optical diffraction.

sample from a transducer, which may be mounted on a short delay rod, to an absorbing termination which prevents standing waves being produced. A parallel beam of monochromatic light is passed through the liquid and focussed on a detector. It is necessary to distinguish between two processes by which the light can be diffracted by the longitudinal elastic wave, Raman—Nath scattering and diffraction at the Bragg angle.

At low frequencies, and with a narrow beam of elastic waves, Raman—Nath scattering predominates. This is the phenomenon observed by Debye and Sears, and Lucas and Biquard, and occurs when the light is incident normally on the beam of longitudinal waves and the wavefronts are parallel to the light beam. The diffracted light appears as a series of equally spaced lines on either side of the primary light beam. Variation of the angle of incidence away from the normal by a small amount results in changes in the relative intensities of the diffracted lines with little change in the angular spacing. A theoretical explanation of this phenomenon was given in a series of papers by Raman and Nath (1935). If the motion of the longitudinal wave is neglected, the successive compressions and rarefractions will result in a periodic variation of the refractive index. A plane wavefront in a light beam entering the medium parallel with the wavefronts of the longitudinal wave will undergo a periodic change of phase across the width of the beam. This results in a "corrugated" wavefront as the beam emerges from the medium. The intensity of the rays of light reaching a distant screen can be shown to have a maximum value at angles θ with the incident beam, where θ is given by

$$\sin\theta = \pm \frac{n\lambda_{opt}}{\lambda_L} \qquad (4.19)$$

where n is the order of the diffracted beam, λ_{opt} is the optical wavelength and λ_L is the wavelength of the longitudinal waves. The diffraction pattern is thus a series of fringes spaced symmetrically about the primary beam. Measurement of the angle θ enables λ_L to be determined from which the velocity of the longitudinal wave can be calculated. Raman and Nath (1936) showed that the effect of the motion of the longitudinal wave is to alter the frequency of the diffracted light by integral multiples of the frequency of the longitudinal wave, f_L. The frequency of the nth order fringe is then

$$f_n = f_0 + nf_L$$

where f_0 is the frequency of the incident light.

If the width of the beam of longitudinal waves is increased, or the frequency increased, the Raman–Nath analysis becomes invalid. When the product of the optical wavelength and the width of the beam of longitudinal waves becomes greater than the square of the longitudinal wavelength ($\lambda_{opt} W > \lambda_L^2$) then the predominant mechanism is that of Bragg diffraction (Klein and Cook, 1967). The diffracted light appears as a single spot and is of maximum intensity when the light is incident on the beam of longitudinal waves at the Bragg angle, given by

$$\sin\theta = \frac{\lambda_{opt}}{2\lambda_L}. \qquad (4.20)$$

Under these conditions light is reflected from adjacent wavefronts with a path difference of $2\lambda_L \sin\theta$ which is equal to the optical wavelength. The angle between the primary and diffracted beam is 2θ and is equal to the scattering angle θ in the Raman–Nath case, as $\sin\theta \ll 1$. The motion of the longitudinal wave again produces a shift in the frequency of the diffracted light by an amount equal to the longitudinal wave frequency.

Figure 4.14 illustrates these two different processes of diffraction

4. HF METHODS FOR MEASURING MECHANICAL PROPERTIES

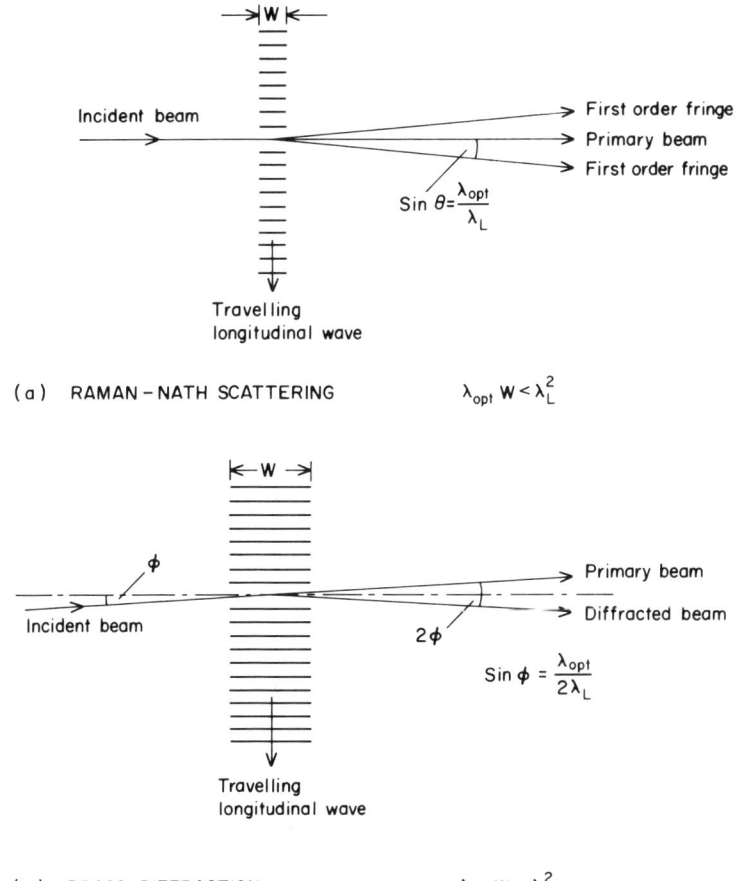

FIG. 4.14. Scattering of light by longitudinal waves; (a) Raman-Nath scattering, (b) Bragg diffraction.

from an elastic wave. The transition between the two processes occurs in liquids at a frequency of about 30 MHz for a beam width in the liquid of the order of 5 mm. The behaviour in the transition region, where both processes are present, is discussed by Willard (1949) and Mayer (1964).

Equations (4.19) and (4.20) must be satisfied within the material in which the longitudinal wave is propagated. If the sides of the sample are parallel to the direction of propagation, the refraction of the light beam on entering and leaving the sample is compensated by the change in the optical wavelength within the sample. Thus the angle of the diffracted beam relative to the primary beam as determined external to the sample can be used in equations (4.19) and (4.20) if λ_{opt} is taken as the optical wavelength in air.

The optical diffraction technique enables the velocity of longitudinal waves to be determined to an accuracy of ±0.1%. The absorption coefficient can be determined by measuring the intensity of the diffracted light at different positions along the beam of longitudinal waves. A laser provides a convenient source of light giving an intense parallel beam. The diffracted beam is then readily visible to the naked eye, eliminating the need to use a photomultiplier as a detector. A photo-diode can be used to monitor the light intensity and modulation of the transducer drive at a frequency of 1 or 3 kHz enables a standard Voltage Standing-wave Ratio (V.S.W.R.) meter to be used to measure the diode output. The temperature of the sample must be controlled if accurate velocity measurements are required. The energy of the longitudinal wave may cause significant heating and the technique is most suitable for use near room temperature.

4.5.2 BRILLOUIN SCATTERING

In the absence of an external source of longitudinal waves, light can be scattered by the naturally occuring elastic waves in a liquid. In this case the scattered light is observed at a large angle, of the order of 90°, from the incident beam. The experimental arrangement for the observation of Brillouin scattering is shown in Fig. 4.15.

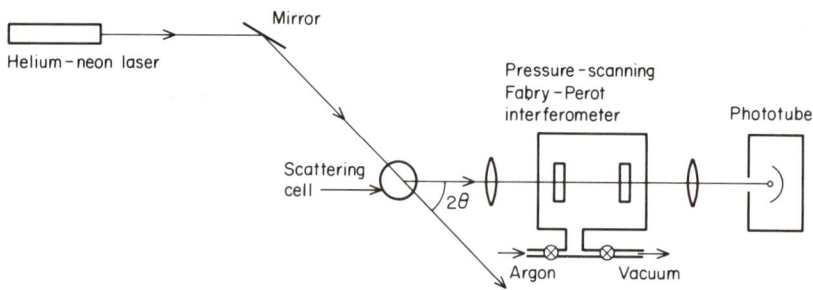

FIG. 4.15. Experimental arrangement used by Fleury and Chiao (1966) to determine velocity and absorption of longitudinal waves in liquids using Brillouin scattering.

4. HF METHODS FOR MEASURING MECHANICAL PROPERTIES

Light will be reflected from a longitudinal wave acting as a diffraction grating if the Bragg condition is satisfied, i.e. $\lambda_{opt} = 2\lambda_L \sin\theta$, where θ is the angle of incidence. The angle between the primary and scattered light is then 2θ. Thus for a chosen scattering angle 2θ, light will be scattered only by longitudinal waves having a wavelength which satisfies the Bragg condition. The scattered light will be frequency shifted by the moving longitudinal wave. Two waves, moving in opposite directions, will satisfy the required conditions. The scattered light thus contains two components, which are shifted equally above and below the exciting frequency by an amount equal to the longitudinal wave frequency. If this frequency shift is measured, the propagation velocity c_L of the wave can be determined, as

$$\Delta f = f_L = \frac{c_L}{\lambda_L} = c_L \frac{2\sin\theta}{\lambda_{opt}} \qquad (4.21)$$

For a scattering angle 2θ of $90°$ and illumination by light of wavelength $0.6\ \mu m$, the frequency shift in liquids having a velocity c_L of the order of $1500\ m\ s^{-1}$ is about 3×10^9 Hz, i.e. a frequency change of less than 0.001%. The observation of frequency changes of this order is possible by using a gas laser as a source of monochromatic light and a Fabry–Perot interferometer as the detector. The technique allows the measurement of velocities at frequencies in the range 1–10 GHz, where conventional methods of generating longitudinal elastic waves are inefficient.

The Fabry–Perot interferometer is a fixed path-length instrument consisting of a pair of partially reflecting plates held parallel a fixed distance apart (Jacquinot, 1960; Born and Wolf, 1964). It is normally used as a scanning instrument, the intensity in the centre of the circular fringe pattern being monitored as the optical separation between the plates is varied causing the diffraction fringe pattern to expand or contract. The effective optical separation between the plates may be altered by changing the pressure of the gas in the system and thus changing the optical wavelength. By evacuating the system and allowing the gas to slowly leak back into it a controlled and linear frequency scan can be obtained. Nitrogen, carbon-dioxide and air have been used as pressurising gases. The light from the centre of the fringe pattern passes through a pinhole onto a photo-multiplier tube. The output from the photo-multiplier is used to drive a chart recorder which gives the scattering spectrum as a plot of intensity versus frequency. Typical experimental systems are described by Chiao and Stoicheff (1964), O'Connor and Schluff (1966), Cummins and Gammon (1966), Pinnow *et al.* (1968a) and Fleury and Boon (1969). An alternative to pressure scanning is to change the plate separation directly by using a piezo-electric material as a mount for one of the plates, or as part of the spacer tube separating the plates. The application of a voltage to the piezo-electric material causes a change in its length which varies the path length in the interferometer (Mielenz *et al.* 1964). Piezo-electric scanning offers the possibility of rapid scanning with the direct display of the spectrum on an oscilloscope. An interferometer using spherical, instead of plane, reflecting surfaces and which incorporates piezo-electric scanning is described by Hercher (1968). The spherical-mirror interferometer has the disadvantage of requiring a fixed mirror separation, and therefore a fixed free spectral range, but has the advantages of higher light gathering power, easier alignment and higher finesse (the ratio of spectral range to bandwidth).

A typical scattering spectrum is shown in Fig. 4.16 and has the two Brillouin peaks symmetrically displaced from the central (unshifted) peak at the exciting frequency. The central peak is due to Rayleigh scattering from density fluctuations. In order to specify the density at any point in a material, two independent thermodynamic variables are required; suitable variables are pressure and entropy. Local density fluctuations arising from pressure changes at constant

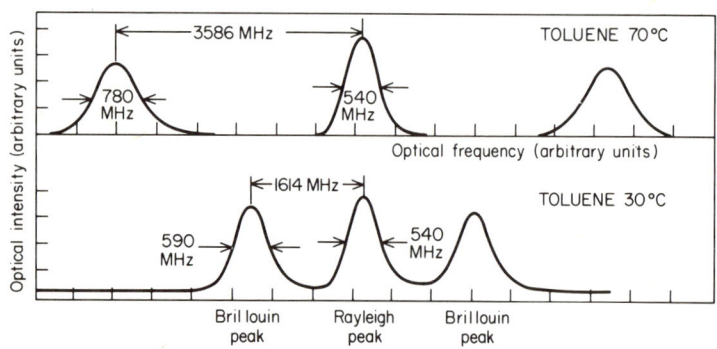

FIG. 4.16. Frequency spectrum of scattered light from toluene at temperatures of 30°C and 70°C (Fleury and Chiao, 1966).

entropy propagate in the liquid as a travelling longitudinal elastic wave. However, fluctuations in density, which are associated with entropy changes at constant pressure, diffuse through the liquid, rather than propagating with a well-defined velocity. Such fluctuations scatter light without any change in frequency and produce the central Rayleigh peak of the spectrum.

Scattering from dust particles (Tyndall scattering) will dominate the central peak unless precautions are taken to eliminate all traces of dust and solid impurities. Distillation of the sample directly into the cleaned optical cell, or filtration through a closed loop containing a suitably fine filter have been found to be satisfactory methods of obtaining a clean sample.

The amplitudes of the Brillouin and Rayleigh peaks are related by the specific heats at constant pressure and constant volume. The ratio of the intensities of the central peak, I_C, to the intensity of the combined Brillouin peaks, $2I_B$, is called the Landau—Placzek ratio (Landau and Lifshitz, 1969) and is given by

$$\frac{I_C}{2I_B} = \frac{C_p - C_v}{C_v} = \frac{\beta_T - \beta_s}{\beta_s} \quad (4.22)$$

where C_p and C_v are the specific heats at constant pressure and constant volume, and β_T and β_s are the isothermal and adiabatic compressibilities. When allowance is made for the dispersion in the velocity of propagation of the longitudinal waves at the frequency of the Brillouin peaks, agreement is found between equation (4.22) and the experimental measurements of the light intensities (Cummins and Gammon, 1966; Mountain, 1966). The intensities are determined from the areas of the peaks in the spectrum. The width of the peaks is related to the attenuation of the density fluctuations and is also dependent on the resolution of the measuring system and the finite bandwidth of the exciting light beam. An analysis of the effect of the system resolution on the width and shape of the peaks is given by Leidecker and La Macchia, (1968).

A damped elastic wave, or a density fluctuation of finite lifetime, has a Fourier spectrum containing a spread of frequencies over a finite bandwidth; a spectrum containing a single frequency component is obtained only for an undamped continuous wave, of constant amplitude. The width of the peaks in the scattering spectrum can be related to the attenuation of the longitudinal waves, in the case of the Brillouin peaks, and to the lifetime of the entropy disturbances producing the central peak. (Montrose et al. 1968). As in the determination of the intensity of the scattered light, the resolution of the measuring system must be considered when determining the width of the peaks. Brillouin scattering measurements thus enable the velocity of longitudinal waves to be measured, at frequencies in the range 1 to 10 GHz, from the position of the peaks in the spectrum, and the absorption coefficient can be estimated from the width of the peaks. Measurements of the intensities of the scattered light enables the isothermal and adiabatic compressibilities to be determined (Pinnow et al. 1968a).

In many liquids, the estimation of the intensities is complicated by the presence of background light scattered by fluctuations in the orientation of the liquid molecules. Consider first the case of scattering without change in the polarisation of the light. The light scattered at an angle of 90° will be polarised perpendicular to the scattering plane (the plane containing the incident and scattered beams): incident light which is polarised parallel to the scattering plane has the polarisation vector parallel to the scattering direction, and thus no scattered light can propagate in this direction. If, however, the plane of polarised light is changed on scattering, due to optical anisotropy in the molecules, a depolarised component will exist in the scattered light, with the plane of polarisation parallel to the scattering plane. Fluctuations in the anisotropy result in the light being scattered with changes in frequency, and the spectrum of the depolarised light contains a broad peak, centred on the incident light frequency, known as the Rayleigh "wing" (Rank et al. 1966; Shapiro and Broida, 1967). Two spectra can then be observed, depending on the direction of polarisation of the incident light, as shown in Fig. 4.17. The depolarised spectrum is due only to orientational fluctuations. The polarised spectrum contains the Rayleigh and Brillouin components as well as a component due to orientational fluctuations. This component, which can be estimated from the amplitude of the depolarised spectrum, must be subtracted from the polarised spectrum in order to obtain the spectrum due to density fluctuations alone. (Starunov et al. 1966; Pinnow et al. 1968b; Cummins and Gammon, 1966).

Detailed studies have been made recently of the depolarised spectrum (Starunov et al. 1967; Stegeman and Stoicheff, 1968,

4. HF METHODS FOR MEASURING MECHANICAL PROPERTIES 105

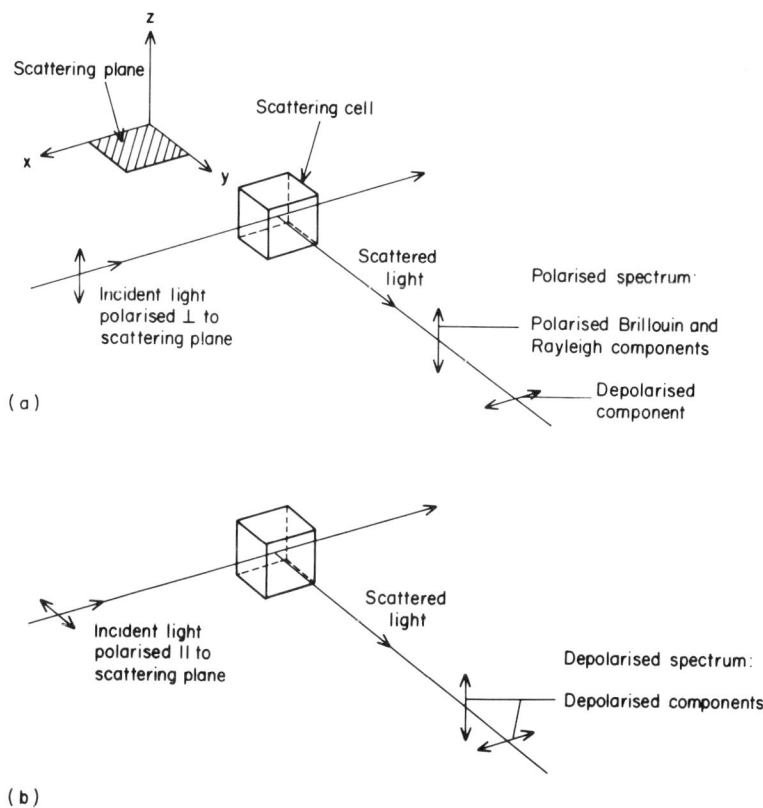

FIG. 4.17. Brillouin scattering, showing the components of the polarised and depolarised spectra.

1973; Enright et al. 1972). In several liquids, this spectrum has been found to consist of a doublet with a spacing of about 1 GHz from the exciting frequency. Leontovich (1941) and Rytov (1958) predicted the presence of such a response, resulting from orientational fluctuations of anisotropic molecules caused by a pair of thermally generated shear waves. The frequency of the shear waves can be determined from the frequency shift of the peaks from the exciting frequency, and the propagation velocity determined using equation (4.21). The theories envisage the liquid as having a complex shear modulus with a single relaxation time τ and a high frequency limiting modulus μ_∞. Typical values, obtained from an analysis of the doublet spectrum, are in the range 10^{-10} to 10^{-11} s for τ, and

10^7 to 10^8 Pa for μ_∞, for liquids having steady flow viscosities of the order of 0.01 Pa s at the temperature of measurement (20°C).

A discrepancy exists in these results, however, in that the product $\tau\mu_\infty$, which should be equal to the viscosity η, is found to be significantly less than η. Barlow et al. (1972a) have suggested that the observed results are more satisfactorily explained in terms of the retardation processes discussed in Chapter 5, Section 5.3. The retardational compliance J_r observed in many liquids gives an effective modulus $G_r = 1/J_r$ in the range 10^7 to 10^8 Pa, comparable with the values of modulus calculated from the light scattering data. The characteristic time τ may be regarded as a retardation time, related to the modulus by a viscosity $\eta_r = \tau G_r$. This viscosity does not contribute to the steady-flow viscosity, and the value of the ratio $\tau G_r/\eta$ is not in general equal to unity.

It is possible therefore that the depolarised component of the light-scattering spectrum can provide information about the shear properties of liquids of low viscosity, at frequencies between 0.5 and 1.5 GHz. These results would then complement the shear wave studies which are confined to liquids of higher viscosity. A further discussion of the viscoelastic, dielectric and light scattering properties of liquids is given in Chapter 6, Section 6.5.

4.5.3 STIMULATED BRILLOUIN SCATTERING

The technique of stimulated, or induced, Brillouin scattering involves focussing a short pulse of high intensity light in the sample to be investigated. A high-power Q-switched ruby laser is used as the source of light. The intensity of the Brillouin component of the scattered light is then sufficiently high so that back-scattered light (a scattering angle of 180°), mixes with the incident light and produces through electrostriction an appreciable pressure wave at the difference frequency between the two light waves. This longitudinal pressure wave will be coherent with the naturally occurring wave causing the scattering at the Brillouin frequency. Thus further scattering will occur, and both the light and longitudinal pressure waves increase in intensity until the incident light energy is depleted. Under these conditions the scattered light can become nearly as intense as the incident light. The back scattered light is reflected into a Fabry—Perot interferometer by a beam splitter, and the interference pattern recorded photographically.

The amplification of the Brillouin component is accompanied by a narrowing of the peak in the spectrum. This reduction makes direct correlation of the width with fluctuation lifetimes impossible and so the absorption coefficient of the longitudinal wave cannot be determined with this technique. A further restriction is that, as amplification will only occur when the incident and scattered light beams are colinear, corresponding to a scattering angle of 180°, the longitudinal wave velocity is determined at one frequency only.

The technique has the advantage that since the measurement is made within the duration of a laser pulse of about 60 ns, the system is insensitive to mechanical vibrations and it is possible to inject the light through the free surface of a sample. A further advantage is the precision with which the position of the narrow Brillouin Peak can be determined. In contrast to a system line-width which may be as high as 1 GHz in a continuous Brillouin scattering experiment, a line-width of as low as 150 MHz can be obtained as a consequence of the gain narrowing of the spectrum peak. Bucaro and Carome (1967), and Goldblatt and Litovitz (1967) have shown that if precautions are taken to ensure that the laser output consists of a single mode, the velocity of longitudinal waves at a frequency around 5 GHz can be obtained to an accuracy of the order of ±0.5%.

5. The Viscoelastic Properties of Supercooled Liquids

The use of the high frequency techniques described in Chapter 4 has shown the existence of viscoelastic behaviour in many low molecular weight liquids. First observed in glycerol (Piccirelli and Litovitz, 1957) and several other associated liquids (Meister *et al.* 1960), similar behaviour is also found to occur in mineral oils (Barlow and Lamb, 1959) and a large number of organic liquids which can be supercooled (Barlow *et al.* 1967a, 1967b).

The limited experimental frequency range requires that temperature be used as an additional variable if the complete relaxation region is to be observed. Data obtained at several temperatures may then be combined to give a composite curve, or "Master" curve, representing the behaviour at a chosen temperature. The method of reduced variables (Section 3.6) enables this process to be carried out if the variation with temperature of the viscosity and the high frequency limiting shear modulus G_∞ are both known.

Over a restricted temperature range, equation (2.6) is a satisfactory representation of the variation of viscosity with temperature for many liquids, with the exponent m having a value of 3 or 4. However, over a wider temperature range, the equation based on the free volume concept, equation (2.8), gives a more satisfactory fit to the data. Extrapolation down to the region of the glass transition temperature, where direct measurement of viscosity is difficult, can then be carried out with confidence, as discussed in Section 2.3.4.

The measurement of the temperature dependence of G_∞ is much more difficult, and requires the use of experimental techniques using shear waves at frequencies of 1000 MHz and above, in order to reach the region where $\omega\tau \gg 1$ and the liquid is behaving elastically. The measurement of G_∞, and its temperature dependence, is discussed in the next section.

5.1 The High Frequency Limiting Elastic Modulus, G_∞

Figure 5.1 shows the variation with temperature of the quantity $R_L{}^2/\rho$ for sec-butyl benzene at several frequencies (Barlow et al. 1967a). These results are typical of the behaviour of many liquids. In the region where $\omega\tau \gg 1$, the reactive component of the shear mechanical impedance, X_L, is much smaller than the real component R_L. Hence from equation (3.7) the real component of the complex shear modulus is approximately given by

$$G'(\omega) \approx R_L{}^2/\rho.$$

The low temperature region of Fig. 5.1 is thus in effect a plot of $G'(\omega)$ against temperature. Further, when $G'(\omega)$ at a given temperature becomes independent of frequency the value is, by definition, G_∞. The variation of G_∞ with temperature is then given by the extreme low temperature region of the curve, where the value of $R_L{}^2/\rho$

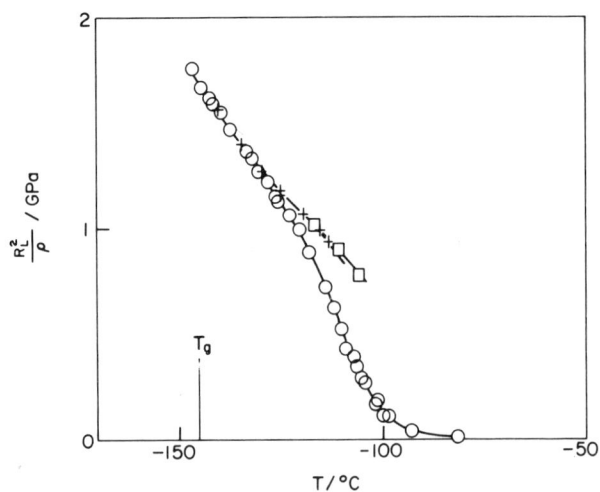

FIG. 5.1. Variation of the quantity $R_L{}^2/\rho$ with temperature at frequencies of 30 MHz (○), 450 MHz (+) and 1000 MHz (□) for sec-butyl benzene (Barlow et al. 1967a).

becomes independent of the frequency of measurement. In order to make a reliable extrapolation of G_∞ into the relaxation region at higher temperatures, it is desirable that the variation of G_∞ with temperature be known over as large a temperature range as possible. The lower temperature limit is the glass transition temperature, T_g. Below this temperature G_∞ becomes sensibly independent of the temperature, as the liquid can not attain an equilibrium state during the course of a normal experiment and remains essentially "frozen" in the state corresponding to T_g. It is therefore apparent that the use of frequencies of the order of 1000 MHz is essential to ensure that the variation of G_∞ with temperature can be observed over a sufficiently extensive region.

The value of G_∞ does not decrease in a linear manner with increasing temperature, although a linear extrapolation has been used by some investigators (Meister et al. 1960). With many liquids, such an extrapolation leads to unrealistic low values of G_∞ in the relaxation region. Rather, the decrease in value becomes less rapid at the higher temperatures, and the limiting compliance J_∞, the reciprocal of G_∞, is found to vary linearly with temperature. Figure 5.2

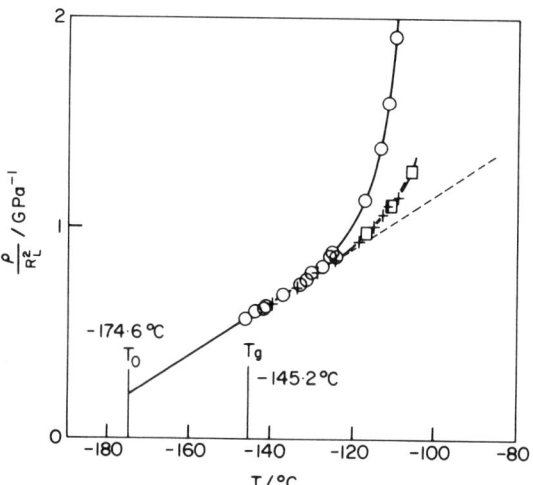

FIG. 5.2. Variation of the quantity ρ/R_L^2 with temperature for sec-butyl benzene. The dashed line shows the linear variation of $1/G_\infty$ with temperature (○, 30 MHz; +, 450 MHz; □, 1000 MHz) (Barlow et al. 1967a).

TABLE 5.1 Values of the limiting elastic modulus G_∞ and the parameter c of equation (5.1)

	$T/°C$	G_∞/GPa	$c/(10^2\ \text{GPa K})^{-1}$
m-bis(m-phenoxyphenoxy) benzene[a]	−25	1.1	0.44
bis(m-(m-phenoxyphenoxy) phenyl)ether[a]	−19	1.1	0.75
boron trioxide[b]	+260	6.0	0.1
1,3 butanediol[c]	−99	2.3	0.30
sec-butylbenzene[c]	−144	1.7	1.4
iso-butylbromide[d]	−160*	1.5	−
sec-butylcyclohexane[e]	−150	2.0	1.8
castor oil[a]	−66	0.9	0.5
di(n-butyl)phthalate[c]	−97	1.5	2.0
di(isobutyl)phthalate[c]	−85	1.5	1.3
di(2-ethylhexyl)phthalate[c]	−89	1.2	1.7
di-octyladipate[a]	−103	1.1	2.1
di(iso-octyl)adipate[a]	−103	1.4	1.9
di(iso-octyl)azelate[a]	−103	1.1	1.0
di-octylsebacate[a]	−113	1.2	1.0
glycerol[e]	−87	5.0	0.18
1,2,6-hexanetriol[f]	−50*	2.6	−
heptamethylnonane[g]	−123	1.3	1.8
2-methyl-2,4-pentanediol[f]	−50*	0.9	−
mineral oil (h.v.i.)[h]	−70*	1.1	2.4
organic glasses[i]	+15	1.3−1.6	−
2-phenylethylchloride[e]	−121	1.6	1.1
3-phenylpropanol[e]	−97	1.5	1.1
3-phenylpropylchloride[e]	−124	1.7	1.4
polybutene, M.Wt. = 448[j]	−40*	0.81	1.7
polybutene, M.Wt. = 2700[j]	−40*	1.8	0.63
poly(n-butylacrylate) M.Wt. = 3.0×10^4[k]	−40*	0.67	2.8
poly(ethylacrylate) M.Wt. = 3.1×10^4[k]	−10*	0.82	1.3
poly(phenylether) (mixed isomeric 4 ring)[a]	−34	1.1	1.4
poly(phenylether) (mixed isomeric 5 ring)[a]	−23	1.2	0.51
1,2 propanediol[l]	−38*	1.1	−
1,5 propanediol[m]	−30*	0.35	−
n-propylbenzene[e]	−148	1.3	1.9

(cont.)

TABLE 5.1 Values of the limiting elastic modulus G_∞ and the parameter c of equation (5.1) (*cont.*)

	$T/^\circ C$	G_∞/GPa	$c/(10^2 \text{ GPa K})^{-1}$
iso-propylbenzene[e]	−151	1.6	1.5
squalane[a]	−109	1.4	1.5
squalene[a]	−108	1.7	1.8
tetra(2-ethylhexyl)silicate[n]	−124	1.6	1.8
tri(β-chloroethyl)phosphate[n]	− 91	2.2	0.7
tris(2-ethylhexyl)phosphate[n]	−113	1.4	1.5
tri-tolylphosphate (mixed isomers)[a]	− 55	1.3	0.55
tri(o-tolyl)phosphate[a]	− 51	1.3	0.4
tri(m-tolyl)phosphate[a]	− 70	2.2	1.3
white oil[a]	− 61	1.1	1.1
nickel[l]	+ 25*	73.5	−
sodium chloride[l]	+ 25*	12.5	−

[a] Barlow *et al.* 1969b.
[b] Capps *et al.* 1966.
[c] Barlow *et al.* 1967a.
[d] Clark and Litovitz, 1960.
[e] Davies *et al.* 1973.
[f] Meister *et al.* 1960.
[g] Hutton and Phillips, 1969.
[h] Barlow and Lamb, 1959.
[i] Crawford, 1956.
[j] Barlow *et al.* 1967c.
[k] Barlow *et al.* 1969a.
[l] Litovitz and Davies, 1965.
[m] Kono *et al.* 1966b.
[n] Barlow *et al.* 1967b.

shows the data of Figure 5.1 plotted as ρ/R_L^2 against temperature, and shows the linear increase of J_∞ with temperature in the low temperature region. This behaviour has been found for many liquids, and may be described by the equation

$$J_\infty = \frac{1}{G_\infty} = \frac{1}{G_0} + c(T - T_0) \qquad (5.1)$$

where T_0 is the reference temperature of the free volume equation, $1/G_0$ is the value at T_0, obtained by continuing the straight line to lower temperatures, and the constant c is obtained from the slope of the line. The extrapolation of the linear region to higher temperatures, shown by the dotted line, may then be used to give the variation of G_∞ with temperature in the relaxation region, for use in normalizing the data according to the method of reduced variables.

Table 5.1 lists the measured values of G_∞ for a wide range of supercooled liquids, polymers and, for the purpose of comparison, some solids. Where possible, the value of G_∞ is given at the glass transition temperature T_g. When the temperature quoted differs from T_g by more than 5° it is marked by an asterisk (*); in these cases the G_∞ values will be somewhat less than the value at T_g.

When referred to the glass transition temperature, so that all the materials are in a similar thermodynamic state, the values of G_∞ for nearly all organic supercooled liquids fall within a very narrow range, from 0.9 to 5 GPa, with the majority of the values lying between 1.1 and 1.5 GPa. This narrow range of values is remarkable, when the wide range of chemical types is considered, and implies that the intermolecular forces are largely independent of the structure of the molecules.

Values of G_∞ of the order of 1 GPa are also found in inorganic glasses, and in many polymers in the neighbourhood of the glass transition temperature. In contrast, most metals and crystalline solids have values of shear modulus between one and two orders of magnitude greater than the liquid modulus, as shown by the last entries in Table 5.1. This difference may be attributed to the presence of long range order and strong intermolecular forces, resulting from the closer molecular spacings in these materials, which are absent in the supercooled liquids and polymers.

For those materials where G_∞ is found to vary with temperature in accordance with equation (5.1), the value of the parameter c is also given in Table 5.1. When referred to the glass transition temperature this equation becomes

$$J_\infty = 1/G_\infty = 1/[G_\infty]_{T_g} + c(T - T_g) \qquad (5.2)$$

where $[G_\infty]_{T_g}$ is the value of G_∞ at T_g, as tabulated in Table 5.1 for the majority of the entries.

The linear dependence upon temperature of the compliance represented by these equations, proposed by Barlow *et al.* (1967a), is at variance with the assumptions made by other workers. For several associated liquids, Meister *et al.* (1960) used a linear variation of G_∞ with temperature, and an exponential decrease of G_∞ with temperature has been assumed in molten zinc chloride (Gruber and Litovitz, 1964) and iso-butyl bromide (Clark and Litovitz, 1960). Hutton and Phillips (1969), in a detailed analysis of measurements on a pure hydrocarbon, heptamethylnonane, suggest that the linear variation of the compliance with temperature is limited to a maximum temperature of about $T_g + 30K$. Thereafter, at higher temperatures, the compliance varies less rapidly with temperature. A variation of this type has also been used by Kono et al. (1966b) in analysing the results of measurements of 1,5-pentanediol.

In the case of the associated liquids, experimental data are available over a very limited temperature range, at temperatures substantially above T_g, and the variation of G_∞ with temperature is smaller than for the majority of the non-associated liquids. It is therefore possible that the results could be described equally well by equation (5.1) or by a linear variation. Matheson (1971a, p. 153) has shown that this is the case for glycerol and also for boron trioxide. In all cases, the variation of G_∞ with temperature which gives a satisfactory reduction to a master curve of data obtained in the relaxation region at differing temperatures and frequencies is taken to be the correct one. This implies that the principle of time—temperature superposition is valid over the complete range of experimental conditions, but in order to test this validity, reliable values of the limiting high frequency shear modulus are needed. This conflict can only be solved, as pointed out by Hutton and Phillips (1969), by actual measurements at higher frequencies than are at present available. Until such measurements are possible, measurements on a large number of liquids indicate that G_∞ varies with temperature in accordance with equation (5.1).

This decrease in the value of G_∞ with increasing temperature is in sharp contrast to the "rubbery" or "entropy" modulus characteristic of rubbers and high molecular weight polymers at temperatures well above T_g and at longer times (Ferry, 1970, Chap. 11). This modulus, with a typical value of 10^5 Pa, is associated with the motion of long chain molecules and is proportional to the absolute temperature. The expression for the variation of this type of modulus with temperature, $\beta = G(T)/G(T_0) = T\rho/T_0\rho_0$, is not applicable to non-polymers or to the high frequency behaviour of polymers.

The temperature dependence of G_∞ for supercooled liquids is given by $(1/G_\infty)(\partial G_\infty/\partial T) = -c[G_\infty]_{T_g}$ at temperatures near T_g, and has a value in the range -0.5 to -3×10^{-2} K^{-1}. This is roughly an order of magnitude greater than the coefficient for a typical solid; in aluminium the value of $(1/G_\infty)(\partial G_\infty/\partial T)$ is -0.72×10^{-3} K^{-1}. This difference parallels the increased volume expansion coefficient of a liquid over a solid, and both effects may be attributed to the variation in free volume with temperature in a liquid, over and above the small changes in interatomic spacings which occur in the lattice of a solid.

5.2 Viscoelastic Relaxation Behaviour of Supercooled Liquids

A knowledge of the variation with temperature of both the viscosity and the high frequency limiting shear modulus enables experimental data obtained over a range of temperatures and frequencies to be plotted in normalized form. The measured values of the components of the complex shear impedance, plotted as $R_L/(\rho G_\infty)^{1/2}$ and $X_L/(\rho G_\infty)^{1/2}$ against $\omega\eta/G_\infty$, are shown in Fig. 5.3 for a large number of pure liquids (Lamb, 1967). Within the experimental error, a single curve represents all the $R_L/(\rho G_\infty)^{1/2}$ data, and a second curve represents all the $X_L/(\rho G_\infty)^{1/2}$ data. The behaviour given by these two curves is found in a large number of pure liquids, among which there are wide differences in molecular structure and composition.

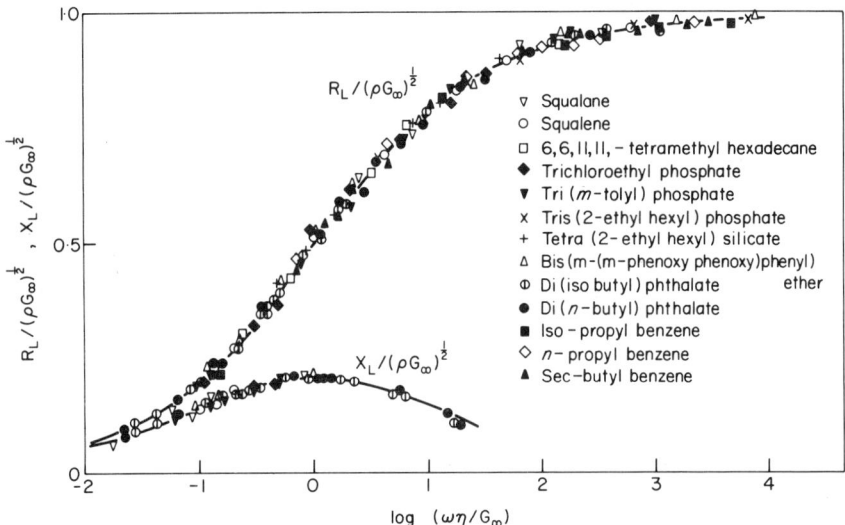

FIG. 5.3. Normalized plots of $R_L/(\rho G_\infty)^{1/2}$ and $X_L/(\rho G_\infty)^{1/2}$ against $\omega\eta/G_\infty$ for a number of supercooled liquids (Lamb, 1967) (reproduced by permission of the Council of the Institution of Mechanical Engineers from their Proceedings).

These liquids include simple derivatives of benzene, phosphate, silicate and phthalate esters, long-chain hydrocarbons and phenyl ethers. Similar behaviour has also been observed in pentachlorodiphenyl by Moore et al. (1969). Mixtures of liquids which are isomers, or which have nearly identical molecular weights also conform to this standard pattern, although large deviations are found for other mixtures. This phenomenon is discussed in a later section.

The viscoelastic relaxation process shown in Fig. 5.3, which appears to be characteristic of pure supercooled liquids, is also found in polymers. In a study of low molecular weight poly(butenes) (Barlow et al. 1967c), a liquid of molecular weight 448, which was nominally pure, having eight repeat units in each molecule, was found to exhibit the same behaviour as the supercooled liquids. This material can hardly be regarded as a polymer, however, and the molecular weight is less than several of the "non-polymeric" liquids (e.g. tetra(2-ethylhexyl) silicate, M = 544). In poly(butenes) of higher molecular weight, up to a maximum of 2700, an additional relaxation process was observed at lower frequencies which can be attributed to the increasing flexibility of the longer molecules. The approach to the limiting modulus at the high frequency end of the relaxation region follows the "pure-liquid" behaviour. The same effect is found in a range of acrylate polymers of molecular weights in the range 4.5×10^3 to 7×10^4 (Barlow et al. 1969a). In these cases, the value of η used in plotting the data on a normalised scale is not the steady flow viscosity, which is determined by the overall motion of the long polymer chains, but a value characteristic of small molecules, reflecting the relative motion of small sections of the polymer chain. The behaviour represented by the curves of Fig. 5.3 may then be regarded as characteristic of the short range interactions of molecules, independent of any long range effects which may be present.

The variation with frequency of the components of the shear mechanical impedance observed in supercooled liquids can be described by an empirical equation (BEL) proposed by Barlow et al. (1967b). The expressions are most conveniently presented in terms of the complex compliance $J^*(j\omega)$. The compliance has the form

$$J^*(j\omega) = \frac{1}{G_\infty} + \frac{1}{j\omega\eta} + 2\left(\frac{1}{j\omega\eta\, G_\infty}\right)^{1/2} \qquad (5.3)$$

or in terms of the components,

$$J'(\omega) - jJ''(\omega) = \left[\frac{1}{G_\infty} + \left(\frac{2}{\omega\eta\, G_\infty}\right)^{1/2}\right] - j\left[\frac{1}{\omega\eta} + \left(\frac{2}{\omega\eta\, G_\infty}\right)^{1/2}\right] \quad (5.4)$$

Comparison with the expressions for the compliance of a Maxwell model, equation (3.15), shows that this equation describes the behaviour of a Maxwell model with the addition of an identical extra term to each component of the compliance. When expressed in normalized form equation (5.3) becomes

$$J^*(j\omega)/J_\infty = 1 + \frac{1}{j\omega\tau_m} + 2\left(\frac{1}{j\omega\tau_m}\right)^{1/2} \quad (5.5)$$

where τ_m is the Maxwell relaxation time, η/G_∞. The derived expressions for the components of the complex rigidity $G^*(j\omega)$ and the components of the shear mechanical impedance $Z_L(j\omega)$ are as follows in normalized form:

$$R_L/(\rho G_\infty)^{1/2} = \frac{(\omega\eta/2G_\infty)^{1/2}[1 + (2\omega\eta/G_\infty)^{1/2}]}{[1 + (\omega\eta/2G_\infty)^{1/2}]^2 + |\omega\eta/2G_\infty} \quad (5.6)$$

$$X_L/(\rho G_\infty)^{1/2} = \frac{(\omega\eta/2G_\infty)^{1/2}}{[1 + (\omega\eta/2G_\infty)^{1/2}]^2 + \omega\eta/2G_\infty} \quad (5.7)$$

$$G'(\omega)/G_\infty = \frac{4(\omega\eta/2G_\infty)^{3/2}[1 + (\omega\eta/2G_\infty)^{1/2}]}{[(1 + (\omega\eta/2G_\infty)^{1/2})^2 + \omega\eta/2G_\infty]^2} \quad (5.8)$$

$$G''(\omega)/G_\infty = \frac{2(\omega\eta/2G_\infty)[1 + (2\omega\eta/G_\infty)^{1/2}]}{[(1 + (\omega\eta/2G_\infty)^{1/2})^2 + \omega\eta/2G_\infty]^2} \quad (5.9)$$

The curves calculated using these equations are shown in Fig. 5.4 for the components of the modulus and the dynamic viscosity ($= G''/\omega$), and in Fig. 5.5 for the components of the shear impedance. The curves of $R_L^2/\rho G_\infty$ versus $\omega\eta/G_\infty$ shown in Fig. 5.6 enable a comparison to be made between a Newtonian liquid, the Maxwell model and the predictions of equation (5.6).

The relaxation process observed in supercooled liquids is considerably broader than the relaxation of a Maxwell liquid. The change in the value of $G'(\omega)$ from 0.1 G_∞ to 0.9 G_∞, for example, which takes about one decade of frequency for the Maxwell model,

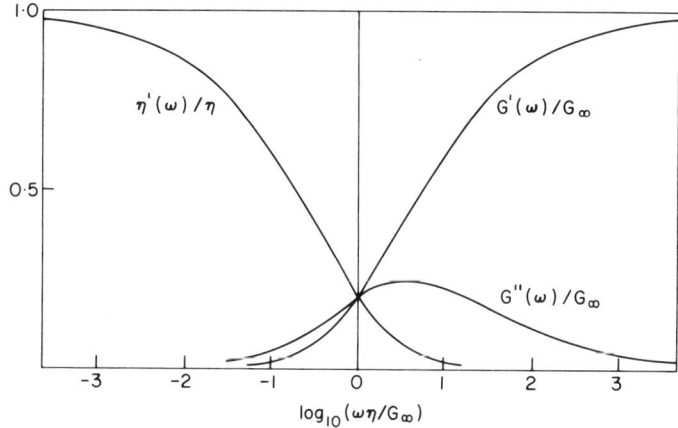

FIG. 5.4. Variation of $G'(\omega)/G_\infty$, $G''(\omega)/G_\infty$ and $\eta'(\omega)/\eta$ against $\log(\omega\eta/G_\infty)$ predicted by equations (5.8) and (5.9) (Barlow et al. 1967b).

takes approximately 3 decades. This behaviour may be expressed in terms of a spectrum of relaxation times, as defined in Section 3.4.1. The spectrum for an assumed distribution of Maxwell elements, derived analytically following the procedure described by Gross (1953), is shown in Fig. 5.7. The dashed vertical line shows the position of the single relaxation time of a Maxwell model.

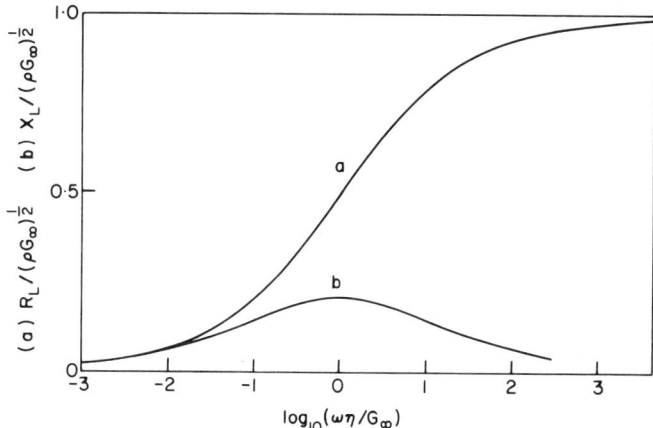

FIG. 5.5. Variation of $R_L/(\rho G_\infty)^{1/2}$ and $X_L/(\rho G_\infty)^{1/2}$ against $\log(\omega\eta/G_\infty)$ predicted by equations (5.6) and (5.7) (Barlow et al. 1967b).

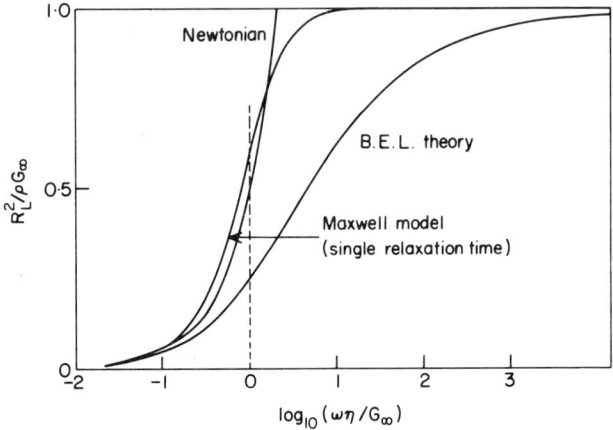

FIG. 5.6. Variation of $R_L{}^2/(\rho G_\infty)$ against $\log(\omega\eta/G_\infty)$ predicted by equation (5.6) compared with the variation for a Maxwell model and a Newtonian liquid (Barlow et al. 1967b).

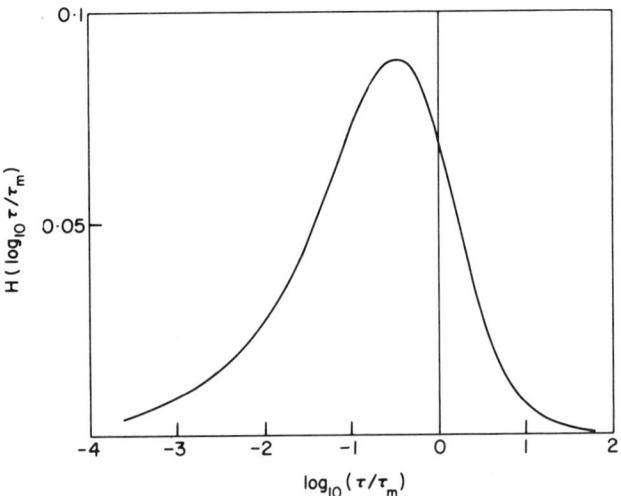

FIG. 5.7. Spectrum of relaxation times for an assumed distribution of Maxwell processes giving a viscoelastic behaviour equivalent to that described by equation (5.3) (from Barlow et al. 1967b).

5. THE VISCOELASTIC PROPERTIES OF SUPERCOOLED LIQUIDS

The close agreement between experimental results and the predictions of the BEL equation, as shown in Fig. 5.3 where the curves are calculated using equations (5.6) and (5.7), suggests that the theory has general validity in its applicability to pure supercooled liquids. A knowledge of the dependence with temperature of the viscosity and limiting shear modulus of a liquid then enables the complete relaxation behaviour to be predicted, within experimental error.

Deviations from the behaviour described by the BEL equation are found in many liquids however, especially in mixtures. Measurements have been made by Barlow *et al.* (1969b) on several binary mixtures of supercooled liquids, the components of which were found to conform to the behaviour described by equation (5.5). At certain concentrations, the behaviour of the mixtures, when plotted in a normalized manner, is identical to that of the pure components. At other concentrations however, significant changes in the width of the relaxation region were observed. The addition of an arbitrary parameter K into equation (5.5) enables the predicted curves to be varied in a systematic manner, the width of the relaxation region increasing as the value of K is increased.

The equation for $J^*(j\omega)$ is then

$$J^*(j\omega)/J_\infty = 1 + \frac{1}{j\omega\tau_m} + 2K\left(\frac{1}{j\omega\tau_m}\right)^{1/2} \qquad (5.10)$$

The families of curves produced by varying K from 0 to 5.0 are shown in Fig. 5.8 for the components of the shear impedance and in Fig. 5.9 for the components of the shear modulus†. $K = 0$ corresponds to the Maxwell model and $K = 1$ corresponds to the original BEL equation (5.5). It was found possible to fit the experimental data for all concentrations of the components to the curves predicted by equation (5.14) by the choice of an appropriate value of the parameter K. Figure 5.10 shows the measured values of $R_L/(\rho G_\infty)^{1/2}$ plotted against log $(\omega\eta/G_\infty)$ for three proportions of a mixture of

---

† Note that in the original paper these figures are incorrectly captioned; Fig. 1 shows the variation of the components of the complex shear modulus, Fig. 2 the variation of the components of the shear mechanical impedance, not the reverse, as captioned.

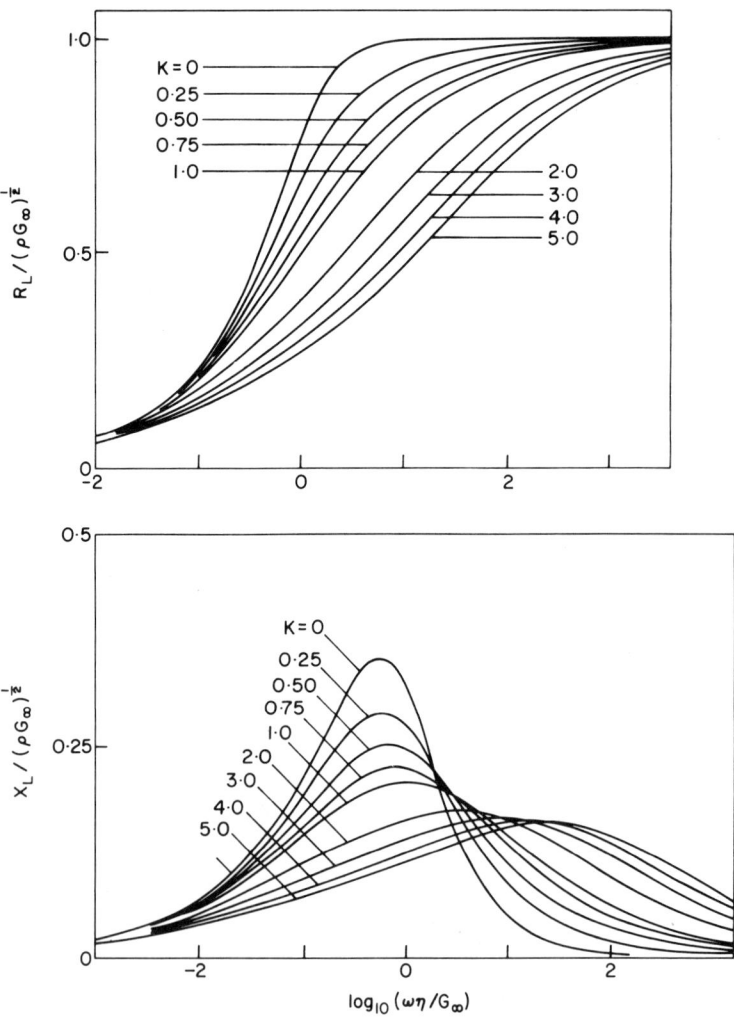

FIG. 5.8. Values of the shear mechanical impedance, calculated from equation (5.10) for different values of the parameter K (Barlow et al. 1969b).

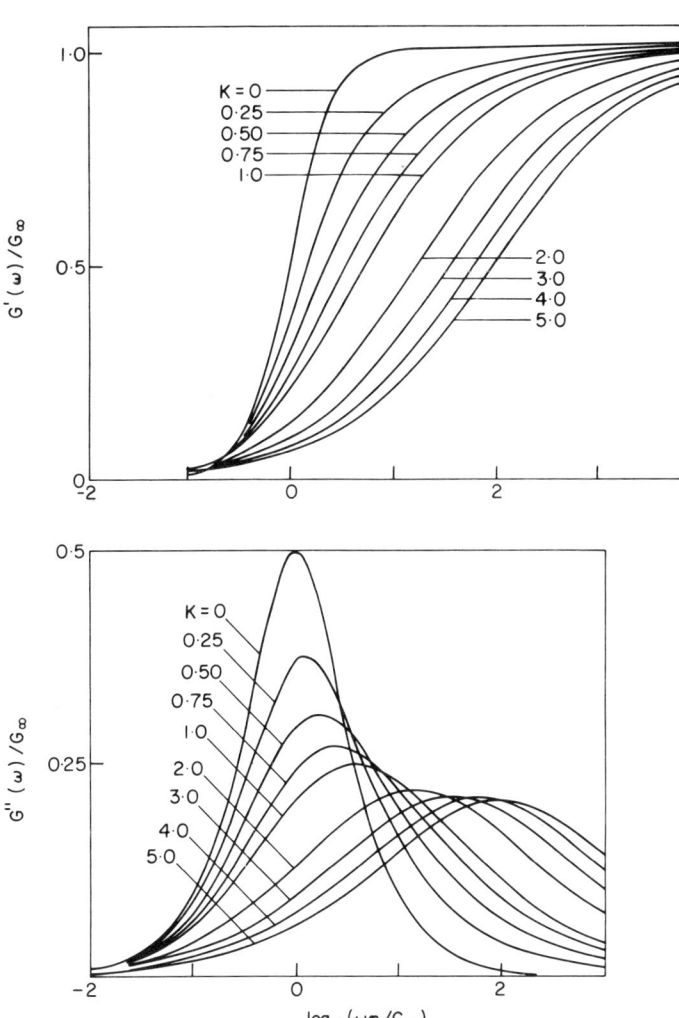

FIG. 5.9. Values of the shear modulus calculated from equation (5.10) for different values of the parameter K (Barlow et al. 1969b).

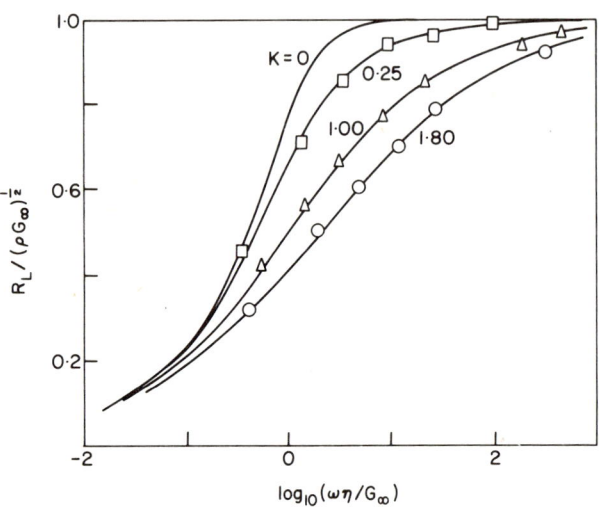

FIG. 5.10. Values of $R_L/(\rho G_\infty)^{1/2}$ plotted against $\log(\omega\eta/G_\infty)$ for three proportions of the mixture $(1-x)$tri(m-tolyl)phosphate + (x)di(n-butyl)phthalate. The curves are calculated for the appropriate value of K (Barlow et al. 1969b).
□, $K = 0.25$, $x = 0.213$
△, $K = 1$, $x = 0.506$ (equimolar)
○, $K = 1.8$, $x = 0.755$

$(1-x)$ parts tri(m-tolyl)phosphate with x parts di(n-butyl)phthalate. The curves are calculated from equation (5.14) with the value of K chosen to give the best fit to the data. The values of K determined in this way are shown in Fig. 5.11 plotted as a function of the mole fraction of the lower molecular weight component. The same systematic variation of K is found for four different binary mixtures. A value of $K = 1$ is found at the equimolar concentration, and also at mole fractions of 0.04 and 0.96; the values correspond to molecular ratios of 1 : 24 and 24 : 1. The minimum value of $K = 0.25$ occurs at a mole fraction of 0.2 (molecular ratio 1 : 4), the maximum value of 1.8 at a mole fraction of 0.8 (molecular ratio 4 : 1). The large variations in K from a value of unity at both very low and very high values of mole fraction suggest that small concentrations of impurity (less than 2%) in a liquid can result in significant changes in the width of the relaxation spectrum. Barlow et al. (1969b) suggest that

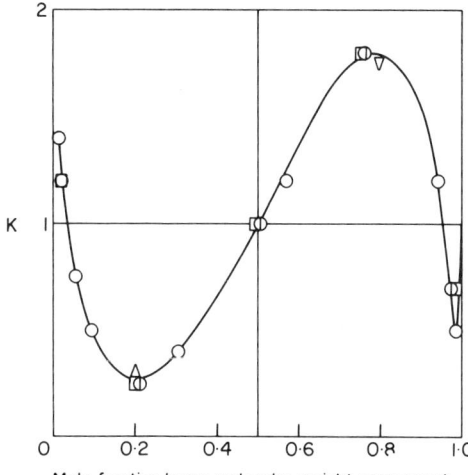

FIG. 5.11. Value of K as a function of mole fraction of the lower molecular weight component for the following mixtures:
○, tri(m-tolyl)phosphate + di(n-butyl)phthalate
□, tetra(2-ethylhexyl)silicate + squalane
△, bis(m-(m-phenoxy phenoxy)phenyl) ether + tri(β-chloroethyl)phosphate
▽, tri(m-tolyl)phosphate + tri(β-chloroethyl)phosphate (Barlow et al. 1969b).

deviations from the behaviour corresponding to $K = 1$ observed in four liquids which were thought originally to be pure, could be due to small amounts of impurity.

Measurements on mixtures of isomers, and liquids of nearly identical molecular weight (squalane (422) and squalene (410); tri(β-chloroethyl)phosphate (285) and di-(n-butyl)phthalate (278)) showed no variation in the value of K away from a value of unity for any concentration. It therefore appears that the components of a binary mixture must differ in molecular weight by an appreciable amount if the behaviour shown in Fig. 5.11 is to occur. Thus, isomers, or impurities of molecular weight close to the value of the main component may be expected to have no effect on the viscoelastic properties of a liquid.

The difference in behaviour observed in castor oil and white oil (Barlow et al. 1969b) can possibly be explained on this basis. Castor oil is a naturally occurring mixture, of which the principle constituents are diglycerides of ricinoleic acid (80—86%), oleic acid (7—9%) and linoleic acid (3—3.5%); the viscoelastic relaxation in castor oil is fitted within experimental error by curves drawn for $K = 2.9$, this being the highest value of K found. White oil consists of a range of saturated hydrocarbons, predominately naphthenic in character. The behaviour of this material can be described by the curves calculated for a value of $K = 1$. If the components present in the white oil do not differ greatly in molecular weight, the behaviour is then consistent with that found in the other mixtures of components of similar molecular weight, and the material behaves as a homogeneous isotropic liquid.

A similar argument can be applied to lubricating oils. Early measurements by Barlow and Lamb (1959) showed the existence of a spectrum of relaxation times in the viscoelastic behaviour of lubricating oils, and an attempt was made to correlate the shape of the spectrum with the chemical composition of the oils. For these oils, G_∞ was found to be independent of temperature. Later work by Hutton (1968) on similar oils, and on a fractionated oil, over a wider temperature range, shows that the spectra cannot be interpreted solely in terms of the hydrocarbon type analysis. Figure 5.12 shows the variation of $R_L/\rho^{1/2}$ and $X_L/\rho^{1/2}$ for one of the oils as a function of effective frequency $f\eta_T/\eta_0$, where η_T is the viscosity at the temperature of measurement, η_0 the value at a reference temperature (30°C) and f is the frequency of measurement. Hutton argues that the general rise in $R_L/\rho^{1/2}$ at frequencies above 10^{12} Hz, corresponding to measurements at temperatures between $-50°$C and $-78°$C at a frequency of 30 MHz, is due to the variation of G_∞ with temperature. The "plateau" in the curve in the region of 10^{13} Hz is present in all the oils measured, but has not been observed in any other liquids. When normalised with respect to G_∞, the data follow fairly closely the behaviour predicted by equation (5.10) with $K = 1$ except in the region in the "plateau". The deviation in this region is thought to be a consequence of the partial crystallization of the oils at the low temperatures used, resulting in a failure of the principle of reduced variables when applied across the phase transition. The constituents of the lubricating oils are hydrocarbons, both aromatic

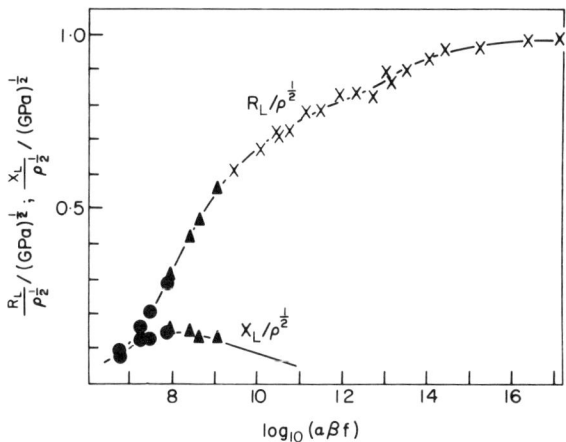

FIG. 5.12. Variation of the components of the shear impedance with effective frequency for a lubricating oil (L.V.I.. 260) (Hutton, 1968).

and saturated, with an average molecular weight of the order of 600, although the spread of molecular weights is typically from below 300 to over 700. The viscoelastic behaviour is close to that of a pure liquid, however, and it may be presumed that the wide range of constituent compounds of differing molecular type prevent the particular structural arrangements which account for the systematic change in spectrum width shown in Fig. 5.11.

Kono et al. (1966a) found that for mixtures of glycerol ($M = 92$) and n-propanol ($M = 60$) the width of the relaxation region increased only slightly as n-propanol was added to glycerol. It is possible, however, that this change is a consequence of some uncertainties in the value of G_∞ and its temperature dependence. Similarly, Slie et al. (1966) found no change in the width of the relaxation region over a wide range of concentrations in mixtures of glycerol and water. The molecular weight of water is only 18, and in this case there is a considerable difference in the molecular weight of the two components. However, if a large proportion of the water were in a structured form, for example the hexagonal-ring form proposed by Davis and Litovitz (1965), the resulting effective molecular weight for water could be much closer to that of glycerol.

Barlow et al. (1969b) tentatively suggest that the behaviour of mixtures of supercooled liquids shown in Fig. 5.11 is a general property of binary liquid mixtures, the components of which differ appreciably in molecular weight, and is determined by the relative numerical distributions of the two types of molecules, independent of the molecular structures.

The common pattern of behaviour observed in many liquids by Barlow, Lamb and co-workers and described by equations (5.3) to (5.9) is at variance with the findings of Meister *et al.* (1960) in a range of associated liquids. In these materials, two different types of distribution of relaxation times were found to be necessary to describe the experimental results. The two distributions used were:

1) A symmetric Gaussian distribution (Wagner, 1913; Yager, 1936) given by:

$$H(\ln\tau/\tau_s) = \frac{b}{(\pi)^{1/2}} \exp - \left[b\ln\left(\frac{\tau}{\tau_s}\right)\right]^2 \quad (5.11)$$

The parameter τ_s determines the position of the spectrum on the time scale, the peak of the logarithmically symmetric curve being at $\tau/\tau_s = 1$. The parameter b determines the width of the spectrum, a value of infinity corresponding to a single relaxation time process.

2) An asymmetric distribution due to Davidson and Cole (1951) given by:

$$H(\ln\tau/\tau_s) = \left(\frac{\sin\beta\pi}{\pi}\right) \left[\frac{\tau/\tau_s}{1 - \tau/\tau_s}\right]^\beta, \; 0 \leq \frac{\tau}{\tau_s} < 1 \quad (5.12)$$

$$H(\ln\tau/\tau_s) = 0, \; 1 \leq \frac{\tau}{\tau_s} < \infty$$

This function has a singularity at the longest relaxation time, τ_s, and the parameter β determines the width of the distribution. At a value $\beta = 1$ the equation reduces to a single relaxation time at $\tau = \tau_s$.

These two distributions are shown in Fig. 5.13 for typical values of b and β.

In 1,3-butanediol, 2-methyl-pentanediol-2,4 and glycerol the symmetrical Gaussian distribution was used, with values of b of 0.46, 0.60 and 0.42 respectively. In 1,2,6-hexanetriol both an asymmetric Cole-Davidson distribution ($\beta = 0.32$) and a single relaxation function were required, the contributions of each process to the modulus

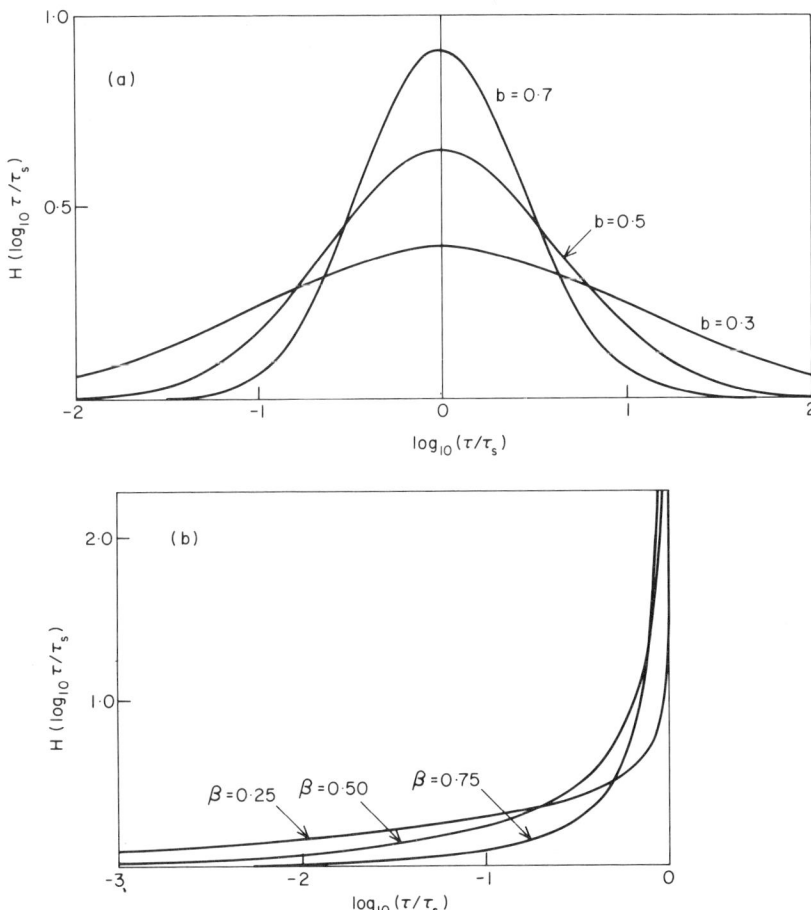

FIG. 5.13. Spectrum of relaxation times $H(\log_{10} \tau/\tau_s)$ for (a) Gaussian distribution, equation (5.11), (b) Davidson–Cole distribution, equation (5.12).

being summed to give the overall response. A linear variation of G_∞ with temperature was assumed for each liquid.

Barlow *et al.* (1969b) have remeasured 1,3-butanediol, and report agreement with the measurements of Meister *et al.* (1960) over the relaxation region. More extensive measurements at low temperatures give values of G_∞ which are thought to be more reliable, and which follow the variation with temperature given by equation (5.1). When normalized with respect to the new G_∞ values, the results are found to conform to the pattern found in other supercooled liquids, as shown in Fig. 5.3. Barlow *et al.* (1969b) therefore suggest that the differences in the behaviour found in the other associated liquids might also be due to uncertainties in the G_∞ data.

A shear relaxation process which can be characterized by a single relaxation time has been observed in two liquids, molten zinc chloride and molten boron trioxide. The viscosity and viscoelastic behaviour of molten zinc chloride have been measured by Gruber and Litovitz (1964). This material is unusual in that the viscosity varies as an exponential function of the absolute temperature — Arrhenius behaviour — as given by equation (2.3), up to much higher viscosities than is the case for organic supercooled liquids. The measurements of Gruber and Litovitz show a linear variation of $\ln(\eta)$ with $1/T$ up to a viscosity of some 3 Pa s in the neighbourhood of the melting point (318°C) although their values of viscosity differ from those given by Mackenzie and Murphy (1960). Measurements of the velocity of shear waves at frequencies of 43, 63, 86 and 118 MHz were made at temperatures in the range 303°C to 358°C. The real component of the complex modulus, $G'(\omega)$, was then determined from the velocity data, assuming a single relaxation process, on the basis of previous compressional relaxation measurements (see Chapter 6). The assumption of an exponential variation of G_∞ with temperature, in the absence of any direct measurements of G_∞, enables the data to be reduced to a single curve which conforms to the predictions of a single relaxation time theory, as shown in Fig. 5.14.

Litovitz and McDuffie (1963), in a comparison of the viscoelastic, ultrasonic and dielectric relaxation behaviour of associated liquids, suggest that the single relaxation time behaviour found in zinc chloride is a consequence of the Arrhenius type viscosity-temperature behaviour. It is argued that in the non-Arrhenius region, where the viscosity is controlled by the available free volume, molecules cannot act individually, but must move in cooperation with their immediate neighbours in order to produce the necessary free volume to allow molecular motion to occur. This cooperative behaviour is regarded as being the origin of the distribution of structural relaxation times observed in liquids exhibiting non-Arrhenius viscosity behaviour.

In liquids which exhibit Arrhenius behaviour, the molecules can move in a non-cooperative manner, independent of their neighbour,

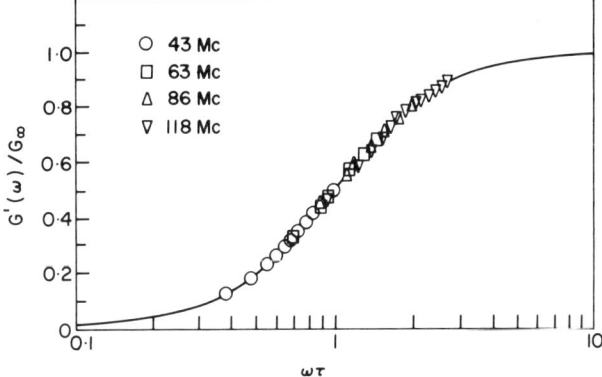

FIG. 5.14. The normalized real part of the shear modulus of zinc chloride versus $\omega\tau$. The solid curve is that predicted by single relaxation time theory (from Gruber and Litovitz, 1964).

in accordance with the rate theory of Eyring (1936) which predicts a viscosity proportional to exp (E/RT), where E is an activation energy. It may be expected that this type of molecular motion would result in a viscoelastic behaviour characterized by a single relaxation time. To test this hypothesis, the viscoelastic behaviour of boron trioxide was studied by Tauke et al. (1968). Over a 600K range, from 1400°C to 800°C this material has a shear viscosity which is Arrhenius. Below 800°C, at which temperature the viscosity is approximately 30 Pa s, the behaviour is non-Arrhenius. Measurements of the viscoelastic behaviour at frequencies between 3 and 45 MHz were made over a temperature range from 600°C to 1000°C, thus encompassing both the Arrhenius and non-Arrhenius regions. The measured data was normalized using G_∞ values determined from ultrasonic relaxation measurements, assuming a constant ratio between the limiting shear and longitudinal moduli. The resulting variation of G_∞ is unusual, having a minimum value some 500K above T_g, then increasing with further increase in temperature. The normalized curves show a relaxation process with a single relaxation time in the Arrhenius region above 800°C, and an increasingly broad distribution of relaxation times with decreasing temperature in the non-Arrhenius region, thus supporting the hypothesis of Litovitz and McDuffie.

The conclusions drawn from these experiments have been criticised on the grounds that these are unusual materials, which do not necessarily behave in a similar manner to organic liquids at lower temperatures. Macedo et al. (1968) suggest that in molten oxides and glasses the temperature dependent spectrum is a consequence of fluctuations in the molecular environment, caused by association fluctuations or bonding differences. The relative size and extent of the fluctuations may be characterized by the parameter Λ, the effective range of the fluctuation. In molten materials a high temperatures Λ is comparable, or less than r_0, the sphere of influence of a molecule. In organic liquids, at the low temperatures where the viscoelastic properties are measured, $\Lambda \gg r_0$, resulting in a distribution of relaxation times which is temperature independent.

5.3 Viscoelastic Retardation in Supercooled Liquids

The viscoelastic behaviour of a large number of liquids and liquid mixtures can be fairly closely represented by the modified BEL equation (5.10), using K as an adjustable parameter. Systematic deviations from the predicted behaviour do occur, however, especially in the region where $\omega \tau_m < 1$. (Barlow and Erginsav, 1972; Davies et al. 1973). A further inadequacy in the equation relates to the behaviour at very low frequencies. The limiting elastic compliance in creep, J_e, can be evaluated from the limiting value of $J'(\omega)$ as $\omega \to 0$ (see Sec. 3.4.2, equation 3.38). From equation (5.10) it can be seen that the low frequency limiting value of $J'(\omega)$ is infinite, implying unlimited elastic strain in creep, a condition which is clearly physically unrealistic. An alternative expression for $J^*(j\omega)$ which has been found particularly successful is based on the Davidson–Cole equation (Davidson and Cole, 1951). This was originally proposed to describe the dielectric relaxation in liquids at low temperatures and in this context is discussed in Chapter 6. Although this equation has been used as a distribution of relaxation times (p. 128), the viscoelastic quantity which is analogous to the complex dielectric permittivity is the complex retardational compliance $J_r^*(j\omega)$ (McCrum et al. 1967), where

$$J_r(j\omega) = J^*(j\omega) - \left[J_\infty + \frac{1}{j\omega\eta} \right]$$

$$= J_r'(\omega) - j J_r''(\omega) \tag{5.13}$$

5. THE VISCOELASTIC PROPERTIES OF SUPERCOOLED LIQUIDS

The viscoelastic form of the Davidson–Cole equation is then

$$J^*(j\omega) = J_\infty \left[1 + \frac{1}{j\omega\tau_m} + \frac{J_r/J_\infty}{(1 + j\omega\tau_r)^\beta}\right] \qquad (5.14)$$

where J_r is the magnitude of the retardational contribution to the compliance and τ_r is the long-time limit of the retardation spectrum. The retardation spectrum $L(\ln\tau/\tau_r)$ (Section 3.4.2) associated with the Davidson–Cole expression for $J_r^*(j\omega)$ then has the form given by equation (5.12), and the form of the distribution is shown in Fig. 5.13. The variation of the components $J_r'(\omega)$ and $J_r''(\omega)$ with frequency is shown in Fig. 5.15. When plotted in parametric form the locus of $J_r^*(j\omega)$ is a skewed arc, with a semicircle as the limiting case when $\beta = 1$ (Fig. 5.16).

The experimental verification of this type of behaviour requires that the equilibrium compliance J_e be measured directly, or inferred from shear impedance data in the region between predominantly Newtonian behaviour and the onset of viscoelastic relaxation. In this region, where $\omega\tau_m \simeq 10^{-2}$, both R_L and X_L are between 10^4 and

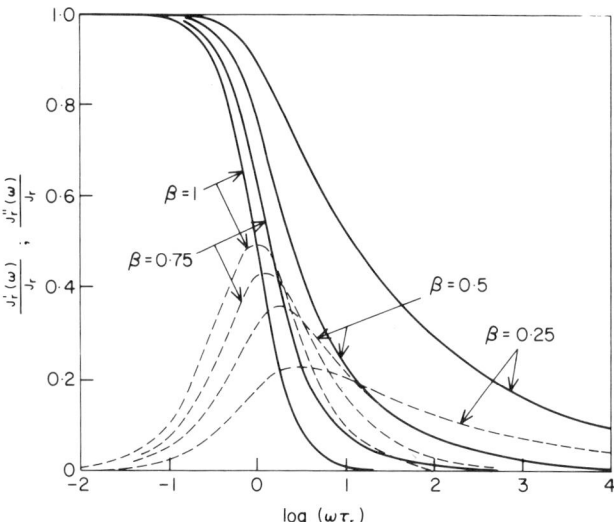

FIG. 5.15. Values of the components of the normalised complex retardational compliance $J_r^*(j\omega)/J_r = J_r'(\omega)/J_r - jJ_r''(\omega)/J_r$ as a function of frequency, calculated from the Davidson–Cole equation (5.14) for several values of β. ———; $J_r'(\omega)/J_r$: – – – –; $J_r''(\omega)/J_r$.

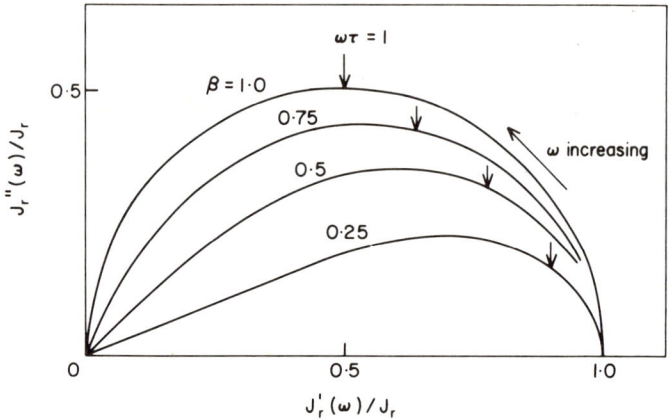

FIG. 5.16. Locus plot of the normalised complex retardational compliance $J_r^*(j\omega)/J_r = J_r'(\omega)/J_r - j(J_r''(\omega)/J_r)$ for the Davidson–Cole equation (5.14) for several values of β.

10^5 N s m^{-3}, and an accuracy of better than $\pm 10^3$ N s m^{-3} in these quantities is required if the value of J_e and the form of the retardation time spectrum are to be deduced from the measured values of the impedance.

An accuracy of this order has been achieved in recent measurements by Barlow and Erginsav (1972) using a shear wave reflection technique operating at 30 MHz. Typical results are shown in Fig. 5.17 for the six liquids measured. In all cases the data can be described within the experimental error by equation (5.14). When plotted to a base of frequency, as in Fig. 5.18, the measured values converge towards the line given by equation (5.5), i.e. $J_r'(\omega) = J_r''(\omega) = (2/\omega \eta J_\infty)^{1/2}$, as $\omega \eta J_\infty$ approaches unity. Thus the differences between the predictions of the Davidson–Cole equation (5.14) and the BEL equation (5.5) are significant only in the region below $\omega \eta J_\infty = 1$. At higher frequencies, equation (5.14) reduces to

$$J^*(j\omega) = J_\infty \left[1 + \frac{1}{j\omega \tau_m} + \frac{J_r/J_\infty}{(j\omega \tau_r)^\beta} \right]$$

$$= J_\infty \left[1 + \frac{1}{j\omega \tau_m} + \frac{1}{(j\omega \tau_m)^\beta} \left\{ \frac{J_r}{J_\infty} \left(\frac{\tau_m}{\tau_r} \right)^\beta \right\} \right] \quad (5.15)$$

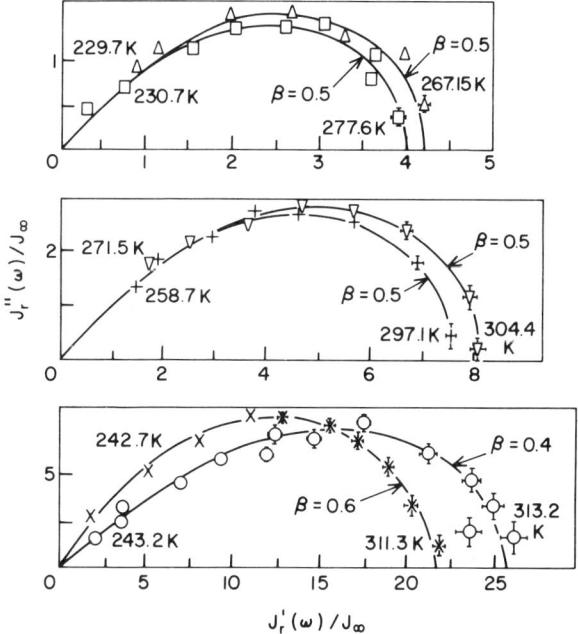

FIG. 5.17. Components of the retardational compliance at 30 MHz.

○ Tri(β-chloroethyl)phosphate ▽ Tri(o-tolyl)phosphate
× Squalane △ Di(n-butyl)phthalate
+ Tri(m-tolyl)phosphate □ Di(iso-butyl)phthalate
(Barlow and Erginsav, 1972).

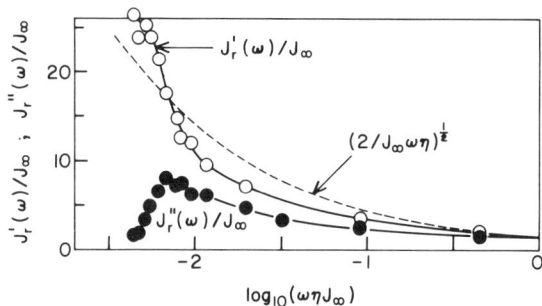

FIG. 5.18. Measured values of $J'(\omega)/J_\infty$ and $J''(\omega)/J_\infty$ for tri(β-chloroethyl) phosphate plotted as a function of $\log_{10}(\omega\eta J_\infty)$ (Barlow and Erginsav, 1972).

This equation is similar to equation (5.10), becoming identical when $\beta = \frac{1}{2}$, $(J_r/J_\infty)(\tau_m/\tau_r)^\beta = 2K$. Equation (5.15) is a useful approximation to the Davidson–Cole equation when the experimental accuracy does not permit evaluation of J_r from the locus plot of $J_r^*(j\omega)$. In this situation the equation can be fitted to the data using β and $(J_r/J_\infty)(\tau_m/\tau_r)^\beta$ as adjustable parameters. This approach has been used by Davies et al. (1973), and their results are included with those of Barlow and Erginsav (1972) in Table 5.2. The value of β is found to lie between 0.38 and 0.76, but is close to 0.5 for the majority of simple organic liquids. The value of J_r/J_∞ varies from near 1, for small rigid molecules, to 85 for a polyglycol with a

TABLE 5.2   Values of retardation parameters in equations (5.14) and (5.15)

	J_r/J_∞	$\dfrac{J_r}{J_\infty}\left(\dfrac{\tau_m}{\tau_r}\right)^\beta$	β
Benzyl benzoate[a]	27	2.0	0.5
s-butyl benzene[b]	–	1.48	0.48
s-butyl cyclohexane[b]	–	1.52	0.53
chloromethylethylene carbonate[c]	1.43	1.6	0.5
di(n-butyl)phthalate[d]	4.2	2.0	0.5
di(iso-butyl)phthalate[d]	4.0	2.0	0.5
eugenol[e]	1	1.41	0.5
glycerol[b]	–	1.41	0.51
2-phenylethylchloride[b]	–	1.29	0.38
3-phenylpropanol[b]	–	2.08	0.39
3-phenylpropylchloride[b]	–	1.20	0.48
poly(glycol) M = 400[f]	17.4	1.72	0.45
poly(glycol) M = 4000[f]	85	10.64	0.76
n-propyl benzene[b]	–	1.67	0.41
iso-propyl benzene[b]	–	1.56	0.43
squalane[d]	21.8	1.24	0.6
tri(β-chloroethyl)phosphate[d]	26.0	1.84	0.4
tri(o-tolyl)phosphate[d]	8.1	2.0	0.5
tri(m-tolyl)phosphate[d]	7.6	2.0	0.5

[a] Barlow and Erginsav, 1974.
[b] Davies et al. 1973.
[c] Ploviec, R., Archiwum Akustyki, to be published.
[d] Barlow and Erginsav, 1972.
[e] Kim, M. G. 1974.
[f] Barlow and Erginsav, to be published.

molecular weight of 4000. The compliance J_r can be associated with the storage of energy under stress through reorientational changes and may be expected to increase as the flexibility or complexity of the molecule increases. The value of J_r is also sensitive to the changes in local packing or molecular orientation which are a function of the concentration in binary mixtures. Barlow and Erginsav (1973) found that the behaviour of a binary mixture which had previously been expressed in terms of equation (5.10), with the variation of K with concentration shown in Fig. 5.11, could also be expressed in terms of equation (5.14). Systematic variation of the parameter β and the ratio of compliances J_r/J_∞ occurred as the concentration was varied, the variations corresponding to the variations in the parameter K. Thus the values of both J_r/J_∞ and β may be affected by small amounts of impurity. Other properties which depend on molecular composition, such as surface tension, viscosity and ultrasonic wave velocity and absorption have also been found to show variations as a function of concentration in mixtures of liquids (Herzfeld and Litovitz, 1959, Ch. XII; Andreae et al. 1965; Solovyev et al. 1968).

In order to use the Davidson–Cole equation to describe experimental results which are obtained over a range of temperatures, it is necessary to assume that the values of β and the ratio J_r/J_∞ are independent of temperature. This is equivalent to the assumption that the form of the relaxation-time spectrum does not change with temperature, as required for the principle of reduced variables to be valid. These assumptions are probably only justified over a limited temperature range, as there is evidence from both viscoelastic and dielectric studies that over a wider range of temperature the distribution becomes narrower with increasing temperature (Kono et al., 1966a, Berberian and Cole, 1968). If the value of J_r can be determined it is possible to determine the ratio τ_r/τ_m as a function of temperature. Barlow and Erginsav (1972, 1973) find that, in all the liquids they studied, the ratio remains substantially constant at low temperatures, but falls rapidly with increasing temperature near the Arrhenius temperature. The retardation time can be expressed in terms of a viscosity $\eta_r = \tau_r/J_r$, which can be compared with the steady flow viscosity $\eta \, (= \tau_m/J_\infty)$. The variation with temperature of these two viscosities is shown in Fig. 5.19 for three binary mixtures. These results are typical of the behaviour of pure liquids and of mixtures, and in the supercooled region both viscosities have a

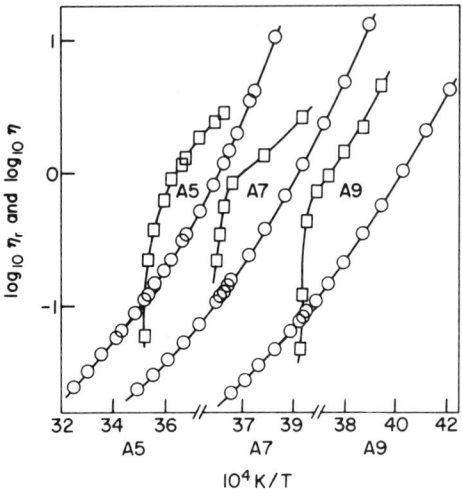

FIG. 5.19. Variation of $\log_{10}\eta$ (○) and $\log_{10}\eta_r$ (□) with $10^4/T$ for three mixtures of tri(m-tolyl)phosphate and di(n-butyl) phthalate (Barlow and Erginsav, 1973).

similar temperature dependence. The quantity $(J_r/J_\infty)(\tau_m/\tau_r)^\beta$ in equation (5.15) is therefore independent of temperature and the predictions of the Davidson–Cole and BEL equations are similar, as the reorientational changes associated with η_r and the compliance J_r are controlled by the same conditions as the viscosity η. Theoretical models based on diffusion mechanisms can give a good description of the behaviour of liquids in this temperature region (see Section 5.5). For ratios of J_r/J_∞ of the order of 10, the retardation time τ_r is typically 10 to 100 times greater than τ_m. As the temperature is increased to the neighbourhood of the Arrhenius temperature the value of η_r falls rapidly and τ_r becomes less than τ_m. Reorientational motions are no longer controlled by the availability of free volume, and τ_r falls to values of 10 to 100 ps which are characteristic of unhindered molecular reorientation times.

The low value of the retardation time in the Arrhenius region possibly explains the single relaxation time (Maxwell) behaviour observed in molten salts. It follows from equation (5.14) that if a liquid exhibits Maxwell behaviour the retardational compliance J_r must equal zero. This conflicts with the view of Goldstein (1969) who concludes that some retarded elastic behaviour necessarily accompanies viscous flow and J_r cannot be equal to zero. The value of J_r/J_∞ is likely to be low in simple liquids such as molten salts, and if τ_r is small compared with the Maxwell relaxation time τ_m the retardational term in equation (5.14) has a negligible contribution to the total compliance in the range of measurement, and the behaviour is indistinguishable from that of a Maxwell liquid.

In di(iso-butyl)phthalate Barlow and Erginsav (1972) observed a second retardation process at temperatures between the Arrhenius temperature T_A and T_K, the temperature where a second change in the viscosity—temperature characteristic occurs (see Section 2.2.4). This process is characterised by a single retardation time, which is longer than the time τ_r in equation (5.14). The additional delayed compliance has a value of approximately $4J_\infty$. This finding is in accordance with the proposals of Davies and Matheson (1966, 1967) relating the viscous behaviour of liquids to the rotational freedom available to molecules. They propose that at temperatures above T_A molecules are able to rotate freely about at least two axes during the interval between translation jumps. In the range between T_A and T_K the reduction in the available free volume allows free rotation about one axis only, and this remaining mode is restricted below T_K. The second retardation process observed in di(iso-butyl)phthalate may then be associated with a molecular motion which becomes progressively restricted in the temperature range between T_A and T_K. Below T_K the remaining rotational mode becomes restricted, giving rise to the retardational behaviour observed in all the liquids studied. At lower temperatures, all reorientational motions are hindered by the lack of available free volume, and only a general diffusional process is observed, which is adequately described by equations (5.5) or (5.15).

A direct measurement of the total recoverable compliance, $J_e = J_r + J_\infty$, has been made in 1,3,5-tri(α-naphthyl)benzene by Plazek and Magill (1966). Measurements of torsional creep and creep recovery were made using a torsional creep instrument in which the rotor is suspended in a magnetic field. Frictional forces are thus reduced to negligible proportions, and the measurement of compliances as low as 10^{-9} Pa^{-1} is possible. It was found possible to reduce data obtained over a range of temperatures near the glass transition temperature to a single curve, implying that the shape of the spectrum of retardation times is independent of temperature. This is consistent with the finding of Barlow and Erginsav (1972) that the quantity β in equation (5.14) is substantially independent of temperature. The reduction of creep data is a much more stringent test than the reduction of relaxational data since the latter is dominated by the contributions of the viscosity and limiting shear compliance. The reduced curve of the recoverable compliance, referred to $64.2°C$ ($T_g = 69°C$) is shown in Fig. 5.20 as a function of time. Barlow and Erginsav have calculated the time dependent behaviour corresponding to the frequency dependent behaviour given by equation (5.14). The resulting curve is indistinguishable from the experimental curve of Fig. 5.20 when the following values are used: $J_\infty = 0.81 \times 10^{-9}$ Pa^{-1}, $J_r = 1.77 \times 10^{-9}$ Pa^{-1}, $\beta = 0.30$ and $\tau_r = 3.55 J_r \eta$ where η is the viscosity of $10^{11.35}$ Pa s at $64.2°C$.

The measurement of the limiting compliance J_∞ and its temperature dependence using a creep technique is complicated by the necessity of making measurements in the immediate neighbourhood of T_g and by the need to extrapolate data to very short times in order to obtain the instantaneous response. It seems likely that a combination of both shear wave and creep measurements would be advantageous in determining the retardation behaviour of super-

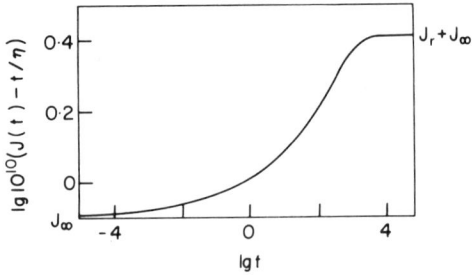

FIG. 5.20. Recoverable compliance of 1,3,5-tri(α-napthyl)benzene at $64.2°C$. Temperature-reduced data of Plazek and Magill (1966). The corresponding variation with frequency is given by $J^*(j\omega) - 1/j\omega\eta = J_\infty + J_r/(1 + j\omega\tau_r)^\beta$ with $J_\infty = 8.1 \times 10^{-10}$ Pa^{-1}, $J_r = 1.77 \times 10^{-9}$ Pa^{-1}, $\beta = 0.3$ and $\tau_r = \eta_r J_r$ where $\eta_r = 3.55\eta = 3.55 \times 10^{11.35}$ Pa s (Barlow and Erginsav, 1972).

cooled liquids. The limiting compliance is more easily measured using high frequency shear wave techniques, while the long time retardational behaviour is conveniently determined by creep and creep recovery measurements.

5.4 Viscoelastic Behaviour of Liquids at High Pressures

Few measurements have been made of the viscoelastic behaviour of liquids under high hydrostatic pressure because of the experimental difficulties involved. The normal-incidence shear-wave reflection technique has been used at pressures up to 1.4 GPa. The usual preliminary measurements with a quartz−air interface cannot be made under pressure, and measurements are first made using a liquid of low and known shear impedance as a reference. If the viscosity of the liquid is sufficiently low, then even at the highest pressure of measurement the shear impedance is given by the expression for a Newtonian liquid, i.e. $Z_L = (j\omega\eta\rho)^{1/2}$. A second series of measurements with the liquid under test in contact with the free end of the quartz delay rod is then made, and from the differences between the two sets of readings, the shear resistance of the test liquid can be calculated. Measurement of the phase change of the signal after reflection at the interface is not normally possible, and so only the real part of the complex impedance, R_L, can be determined. Figure 5.21 shows a typical arrangement of the acoustic system for use under pressure. An O-ring seal separates the test liquid from the liquid backing the transducer, and metal bellows allow for the compression of the liquids under pressure. Measurements of the real part of the shear impedance have been made as a function of pressure, at various temperatures in glycerol (Slie and Madigosky, 1968), di(2-ethylhexyl)phthalate (Hutton and Phillips, 1972) and castor oil (Barlow et al. 1973). Measurements at 30°C have also been made on di(2-ethylhexyl)phthalate and a six-ring polyphenyl ether (bis-(m-(m-phenoxy)phenoxy)phenyl ether) by Barlow et al. (1972b). In all these liquids the limiting modulus G_∞ varies linearly with pressure, the behaviour shown in Fig. 5.22 is typical. The atmospheric pressure value of G_∞ is obtained by extrapolation from measurements made as a function of temperature. In all cases the straight line fitted

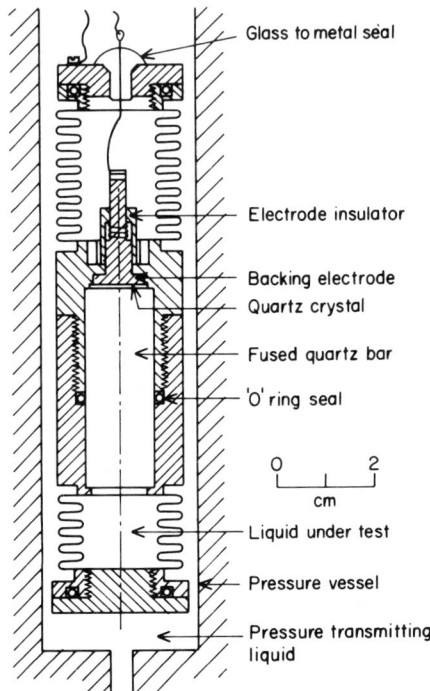

FIG. 5.21. Arrangement of the acoustic system for measurements of mechanical shear resistance under pressure (Barlow et al. 1972b).

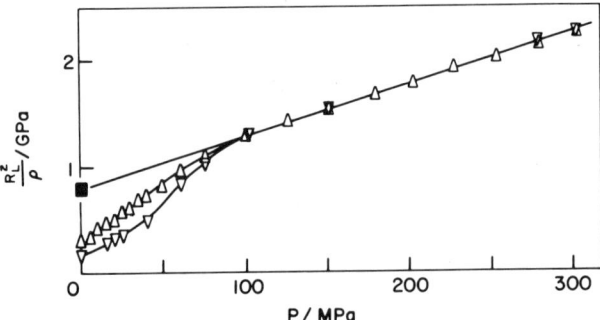

FIG. 5.22. Measured values of $R_L{}^2/\rho$ as a function of hydrostatic pressure for bis-(m-(m-phenoxy phenoxy) phenyl)ether at 30.0°C. ▽, 10 MHz; △, 30 MHz; ■, extrapolated value obtained by variation of temperature at atmospheric pressure (Barlow et al. 1972b).

to the high pressure G_∞ data intersects the ordinate axis at the value of G_∞ estimated from atmospheric pressure measurements.

The rate of increase of G_∞ with pressure shows some dependence on temperature, being greater at low temperatures, as shown in Fig. 5.23 for di(2-ethylhexyl)phthalate. The value of $\partial G_\infty/\partial P$ increases from 1.6 to 2.6 over the temperature range $+30°C$ to $-30°C$, an increase of over 60%. This contrasts with the increase of less than 10% over the same temperature range reported in castor oil, and the constant value assumed for glycerol. At $30°C$, the value of $\partial G_\infty/\partial P$ is about 1.5 for castor oil and di(2-ethylhexyl)phthalate, 3 for glycerol and 5 for the polyphenyl ether. The pressures encountered in elasto-hydrodynamic lubrication are typically greater than 1 GPa. The value of G_∞ at atmospheric pressure is usually less than 1 GPa and thus the pressure coefficient of the modulus may be more important under these conditions than the particular value at

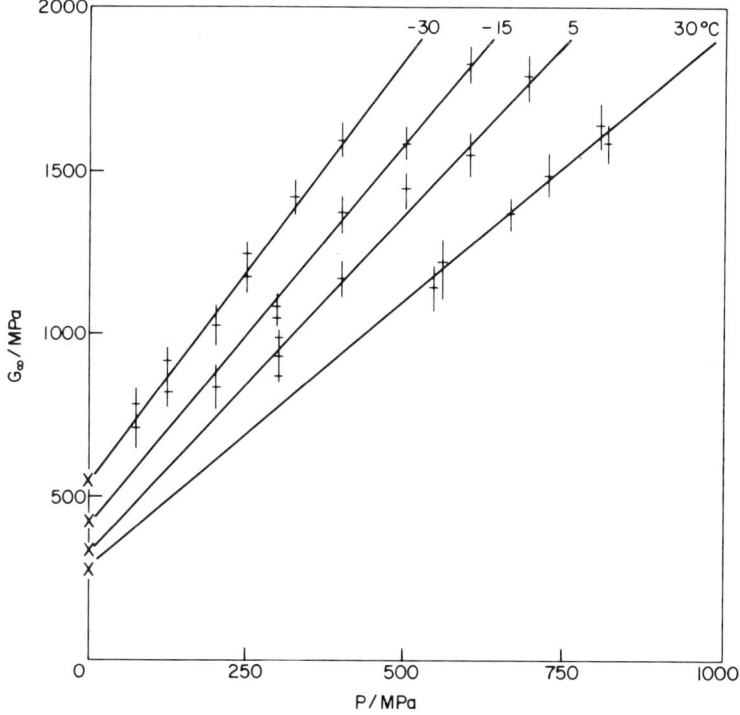

FIG. 5.23. Variation of G_∞ with pressure for di(2-ethylhexyl) phthalate at several temperatures (Hutton and Phillips, 1972).

atmospheric pressure. Hutton and Phillips (1972) suggest that the low value of $\partial G_\infty/\partial P$ found in the synthetic lubricant di(2-ethylhexyl)phthalate may account for some of the differences existing between lubrication theory and the experimental results.

Presentation of the shear impedance data as a function of normalised frequency is possible if the variations of the density and viscosity of the liquid with pressure are known. The results of Slie and Madigosky (1968) on glycerol were found to reduce to the same normalised curve as previous data obtained at atmospheric pressure as a function of temperature. This curve is shown in Fig. 5.24 and can be described by a log-Gaussian distribution of relaxation times, equation (5.11). A similar agreement between data obtained both as functions of pressure and temperature is observed by Barlow *et al.* (1972) although in this case the reduced curve, shown in Fig. 5.25, is described by equation (5.3). The successful reduction of the pressure data to a single curve implies that the shape of the spectrum of relaxation times is independent of pressure as well as temperature. Thus pressure can be used as an equivalent variable to temperature in extending a limited experimental frequency range. The use of

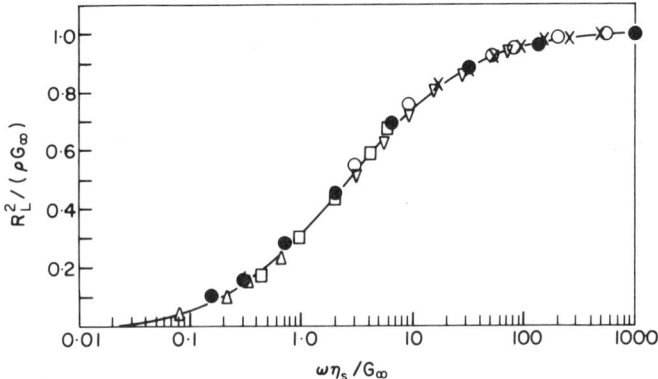

FIG. 5.24. Normalised values of the real component of the shear modulus of glycerol (Slie and Madigosky, 1968).

- ● 85 MHz ⎫ Atmospheric pressure,
- ○ 117 MHz ⎭ Variable temperature.
- △ 15 MHz ⎫ $-16°$C, variable pressure.
- □ 86 MHz ⎭
- ▽ 15 MHz ⎫ $17°$C, variable pressure.
- × 86 MHz ⎭

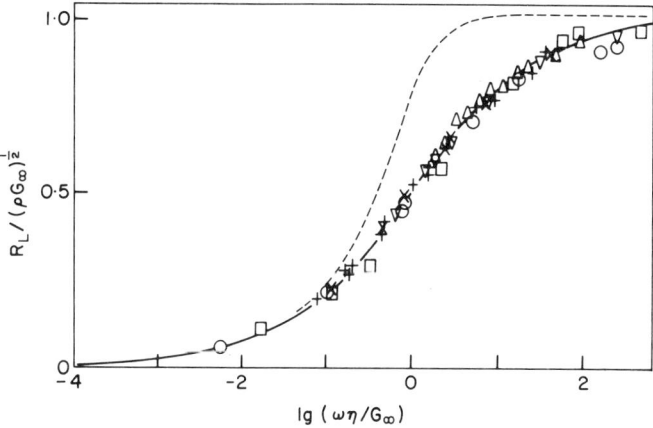

FIG. 5.25. Normalised values of the resistive component of the shear mechanical impedance. Obtained at frequencies in the range 6 to 450 MHz and as a function of temperature at atmospheric pressure;
+, bis-(m-(m-phenoxy phenoxy)phenyl)ether;
×, di(2-ethylhexyl)phthalate.
Obtained at 30.0°C as a function of pressure;
▽, bis-(m-(m-phenoxy phenoxy)phenyl)ether, 10 MHz;
△, bis-(m-(m-phenoxy phenoxy)phenyl)ether, 30 MHz;
○, di(2-ethylhexyl)phthalate, 10 MHz;
□, di(2-ethylhexyl)phthalate, 30 MHz;
The solid line represents the behaviour described by equation (5.3). The broken line shows the theoretical variation for a single relaxation process (Barlow et al. 1972b).

pressure is particularly advantageous in the case of liquids such as silicone oils, which have a high compressibility but which have a low variation of viscosity with temperature.

5.5 Theories of Viscoelastic Behaviour

The various expressions given in this chapter which have been used to describe the viscoelastic behaviour of liquids, i.e. the Gaussian and Davidson—Cole functions and the equations proposed by Barlow, Erginsav and Lamb, are convenient, but arbitrary, mathematical representations of the experimental results. Several attempts have been made to derive similar expressions from fundamental considerations and thus to relate the various parameters to the molecular motions of the liquid.

The origin of the distribution of relaxation times observed in most liquids has been the subject of considerable study. The application of an external constraint to a liquid results in a change in the distribution of the molecular states. In viscoelastic relaxation, an applied strain results in the displacement of molecules from their equilibrium positions. The externally observed stress resulting from these displacements is relieved by molecules moving to new positions by the processes involved in shear flow, until a new equilibrium is established. The rate of return to an equilibrium state is determined by the rate at which the relevant molecular processes can take place.

Frohlich (1958) and Kauzmann (1948) have suggested that differences in the environments of different molecules, due to random thermal fluctuations, would result in a distribution of the activation energies for molecular motion. Different molecules would then achieve equilibrium at different rates resulting in a distribution of relaxation times. An increase in temperature should broaden the relaxation region, as greater differences in local environment are possible. This result is contrary to experiment, as the width of the spectrum is usually found to be independent of temperature over a considerable temperature range, or, in a few cases, to be reduced in width by increasing temperature. Changes in environment caused by the presence of impurity molecules should also result in an increase in the width of the spectrum, if this hypothesis is correct. Again, this is contrary to experiment, as a decrease in width can be produced by the introduction of small amounts of impurity.

The "hole" theory of Eyring (1936) proposes that the flow of liquids takes place by the exchange of the positions of a molecule and a "hole", a molecular defect or vacancy. The theory assumes that the probability of a molecule having sufficient energy to jump into a hole is dependent on temperature only and not on the behaviour of neighbouring molecules. All molecules thus behave in a similar manner and the viscoelastic behaviour can be described by a single relaxation time. The theory predicts a viscosity proportional to exp E/RT where E is an activation energy. The limited results available in liquids having the Arrhenius viscosity behaviour would appear to support this theory (Gruber and Litovitz, 1964; Tauke et al. 1968).

Litovitz and McDuffie (1963) have suggested that the distribution of relaxation times observed in liquids exhibiting non-Arrhenius behaviour is due to the need for molecules to move in a co-operative manner as a consequence of the limited amount of free volume. Their results have already been discussed in some detail.

The concept used by Litovitz and McDuffie of a region of short range order which is continually breaking up and reforming is also central to the theory of Isakovitch and Chaban (1966). They visualise a liquid as made up of regions of short range order interspersed with disordered regions. Diffusion of defects (holes) between these areas results from a disturbance of the equilibrium conditions. The theory predicts a single form of expression for $J_r{}^*(j\omega)$ which is close to the observed behaviour in several liquids. The theory cannot take account of the different widths of spectra that are experimentally observed, however. It also suffers from the disadvantage that the ordered regions are required to remain stable for periods of time which are long compared with the period of the ultrasonic wave. Montrose and Litovitz (1970) point out that the breakup of these clusters at longer times has never been observed experimentally.

The concept of molecular diffusion as a process contributing to structure relaxation is used in several other theories. It is assumed that holes diffuse through the liquid, and relaxation of a molecule occurs instantaneously whenever a hole reaches the site of the molecule. A one-dimensional defect diffusion model due to Glarum (1960a) results in an expression which is in good agreement with the results of dielectric and nuclear magnetic relaxation at short times. The addition of a second relaxation process was necessary to account for the observed behaviour at long times. Phillips et al. (1972) have suggested that the failure of the Glarum theory at long times is because only the nearest defect to a molecular site is considered. Consideration of the effects of the more remote defects results in expressions which are mathematically intractable. Mathematical approximations to the exact solution, which approach the forms of the single-defect diffusion model at short times and yet give acceptable behaviour at long times, have been developed by Phillips. His second-approximation solution results in an expression which is identical to the empirical equation of Barlow et al. (1967b), which has been shown to provide an excellent description of the behaviour of many pure supercooled liquids, particularly at the higher frequencies ($\omega > \eta/G_\infty$).

Montrose and Litovitz (1970) combine the effect on the local order of diffusion over a distance σ, characterised by a time σ^2/D, with a simple rate process characterised by a relaxation time τ_0. The quantity D is a diffusion coefficient, and the two processes are regarded as corresponding to the slow motion of holes by many small steps (diffusion) and the sudden appearance (or disappearance) of a hole produced by a molecular jump. By varying the ratio $\sigma^2/D\tau_0$ a family of curves is generated, similar to that described by the Davidson–Cole equation for various values of β. This theory thus overcomes the disadvantage of the Isakovich and Chaban theory of only a single expression for $J_r^*(j\omega)$, and the predictions are compatible with a wide range of experimental results. For many liquids the ratio of $\sigma^2/D\tau_0$ is found to be close to 4. The predicted behaviour of $J_r^*(j\omega)$ for this value of $\sigma^2/D\tau_0$ is shown in Fig. 5.26, which also shows the predictions of the Davidson–Cole expression for $\beta = 0.5$ (equation 5.14) and the theory of Isakavich and Chaban. The general forms of the curves are seen to be very similar, and when expressed as the complex shear modulus $G^*(j\omega)$ the differences are small, and are confined to the low frequency region where $\omega < \eta/G_\infty$.

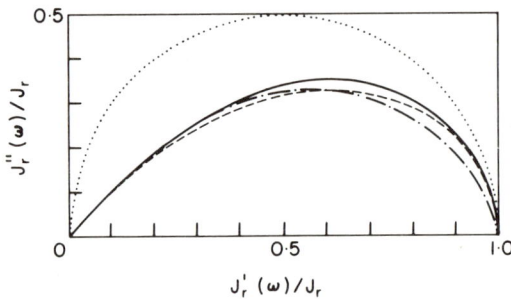

FIG. 5.26. Relative variation of the components of the complex retardational compliance. ——— Davidson–Cole equation with $\beta = 0.5$; —·—·— theory of Isakovich and Chaban (1966); ———— theory of Montrose and Litovitz (1969) with $\sigma^2/D\gamma_0 = 4$; ······· single retardation time (Barlow and Erginsav, 1972).

Thus, several theories which incorporate the concept of the diffusion of a defect are capable, with the appropriate choice of the parameters, of qualitatively describing the observed viscoelastic behaviour of supercooled liquids. In all cases the parameters cannot be predicted from molecular quantities but must be deduced from a comparison between the theoretical predictions and the experimental results. The use of the method of reduced variables, or time–temperature superposition principle, is usually necessary in order to obtain normalised data in a form suitable for comparison with these theoretical predictions. The various theories thus in no way provide any check of the validity of the assumptions inherent in the principle of reduced variables. It is therefore still necessary for measurements to be made over as wide a frequency range as possible in order to obtain reliable data.

6. Structural and Dielectric Relaxation

6.1 Structural Relaxation

The study of the propagation of high frequency (20kHz − 1GHz) longitudinal elastic waves in a material is usually described as "Ultrasonics". The motion of such waves is the same as that of plane sound waves in air, and so the term "acoustic wave" is commonly used to describe longitudinal waves in any material, even though the frequency may be well above the audible range. Other types of mechanical elastic wave, especially surface waves, are also commonly described as acoustic waves when it is necessary to distinguish them from an associated electro-magnetic wave.

The temperature changes associated with a propagating longitudinal wave perturb an existing molecular equilibrium condition, and the use of frequencies up to 1GHz enables the characteristics of very fast reactions to be determined. Such thermally excited equilibria include rotational isomerism, the relaxation of the vibrational specific heat and monomer-dimer reactions in carboxylic acids (Herzfeld and Litovitz, 1959; Lamb, 1965; Beyer and Letcher, 1969; Gooberman, 1969).

The equilibrium state of a liquid can also be perturbed by the change in local pressure associated with a longitudinal wave as molecules move between positions of high and low density. After a sudden pressure change the new equilibrium state is not reached instantaneously. In the simplest case, the final state is approached in an exponential manner, with a time constant τ. The process is termed a relaxation process, as for the time dependent response to a shear stress discussed in Chapter 3, and the time τ is called the relaxation time. The effects of this "structural" relaxation can be described either in terms of a relaxing specific heat, the value decreasing as the frequency of the wave is increased, or by a relaxing bulk modulus, in which case the value increases with increasing frequency. This change in the bulk modulus parallels the change observed in the shear modulus; the high frequency value is characteristic of the glassy state, the smaller low frequency value characteristic of the liquid state.

The intensity of a longitudinal wave decreases exponentially with distance as it passes through a liquid, the energy being dissipated as heat. Two mechanisms provide the "classical" absorption which is present in all liquids. Of these the more important is due to the viscous losses arising from the shearing motion inherent in a plane longitudinal wave. The second source of absorption is due to heat conduction. The sinusoidal pressure variation at any point in the liquid results in a cyclic variation in the local temperature. This occurs in the majority of liquids for which the specific heat at constant pressure C_p exceeds that at constant volume C_v. Heat will flow from the hotter compression regions to the cooler rarefaction regions so that the temperature variation along the wave will tend to be smoothed out, resulting in an energy loss from the wave. With the exception of molten metals the absorption due to heat conduction in liquids is very small compared to that caused by the shear viscosity. The total absorption from these two mechanisms is readily calculated and leads to a constant value of the quantity α/f^2, where α is the amplitude absorption coefficient and f is the frequency of the wave (Herzfeld and Litovitz, 1959, Chapter 1).

However, in most liquids, measurement of the absorption coefficient gives a value for α/f^2 which is much greater than the classical value, indicating the presence of further loss mechanisms. The concept of a volume, or bulk, viscosity associated with the pure compression of a liquid is used to account for the absorption in excess of the classical absorption. The attenuation of the longitudinal wave occurs only when the period of the wave is long compared with the relaxation time τ, i.e. $\omega\tau \ll 1$, and the liquid can readily attain a new equilibrium state. When the period of the wave is short compared with τ, so that $\omega\tau \gg 1$, no change in the state of the liquid is possible during one cycle of the wave, and the attenuation is thus reduced. The increase in the bulk modulus with frequency results in the velocity of propagation of the wave increasing over the relaxation region. The measurement of the absorption and velocity of longitudinal waves, and the calculation of the relaxation parameters is discussed in the following sections. A detailed discussion of structural relaxation in supercooled liquids, its dependence on frequency, temperature and pressure and its relationship to visco-elastic relaxation is given in the later sections of this chapter.

6.2 Propagation of Longitudinal Waves in Liquids

For a progressive plane longitudinal wave propagating in the y direction in an infinite elastic material the wave equation can be shown to be (Hertzfeld and Litovitz, 1959)

$$\frac{\partial^2 p}{\partial t^2} = \left(\frac{K + \frac{4}{3}G}{\rho}\right) \frac{\partial^2 p}{\partial y^2} \quad (6.1)$$

where p is the acoustic pressure, that is, the change in the pressure at any point from the ambient pressure. K is the bulk modulus given by

$$K = V\frac{\partial P}{\partial V} = V\frac{p}{\partial V}.$$

G is the shear modulus and ρ is the equilibrium density. The velocity of propagation of the wave c_L is given by the expression

$$c_L{}^2 \rho = K + \tfrac{4}{3}G$$

or
$$c_L = \left(\frac{K + \frac{4}{3}G}{\rho}\right)^{1/2} \quad (6.2)$$

The quantity $(K + \frac{4}{3}G)$ is called the longitudinal modulus and is denoted by M. The wave is attenuated in propagating through the material and K and G will be complex, frequency dependent, quantities. Thus c_L and M will also be complex and frequency dependent. Denoting complex quantities by an asterisk (*) equation (6.2) may be written

$$M^*(j\omega) = M'(\omega) + jM''(\omega) = K^*(j\omega) + \tfrac{4}{3}G^*(j\omega) =]\, c_L{}^*(j\omega)]^2 \rho$$

$$(6.3)$$

A solution to equation (6.1) for the case of harmonic variations of angular frequency ω is

$$p = p_0 \exp [j\omega(t - y/c_L{}^*(j\omega))] \qquad (6.4)$$

In terms of an attenuation coefficient α and phase velocity $c_L(\omega)$ this becomes

$$p = p_0 \exp [-\alpha y] \exp [j\omega(t - y/c_L(\omega))]$$

where
$$\frac{1}{c_L{}^*(j\omega)} = \left(\frac{1}{c_L(\omega)} - j\frac{\alpha}{\omega}\right) \qquad (6.5)$$

The attenuation coefficient is also a frequency dependent quantity, but is usually denoted simply as α, rather than $\alpha(\omega)$; this convention will be followed here.

The bulk modulus $K^*(j\omega)$ comprises a frequency independent part K_0 plus a frequency dependent relaxational part $K_r{}^*(j\omega) = K'(\omega) + jK''(\omega)$.

Then
$$\left. \begin{array}{c} K^*(j\omega) = K_0 + K'(\omega) + jK''(\omega) \\ \text{and from section 3.1,} \quad G^*(j\omega) = G'(\omega) + jG''(\omega) \end{array} \right\} \qquad (6.6)$$

Then
$$\begin{aligned} M'(\omega) + jM''(\omega) &= [c_L{}^*(j\omega)]^2 \rho \\ &= [K_0 + K'(\omega) + \tfrac{4}{3}G'(\omega)] \\ &\quad + j[K''(\omega) + \tfrac{4}{3}G''(\omega)] \end{aligned}$$

Substituting for $c_L{}^*(j\omega)$ from equation (6.5), and assuming that $(\alpha c_L(\omega)/\omega)^2 \ll 1$ gives the approximate equations

$$M'(\omega) = K_0 + K'(\omega) + \tfrac{4}{3}G'(\omega) = \rho [c_L(\omega)]^2 \qquad (6.7)$$

$$M''(\omega) = K''(\omega) + \tfrac{4}{3}G''(\omega) = 2\rho\alpha [c_L(\omega)]^3/\omega \qquad (6.8)$$

Equations (6.7) and (6.8) enable the components of the modulus $M^*(j\omega)$ to be determined from the measured values of attenuation coefficient α and velocity $c_L(\omega)$.

At low frequencies, where $K_r{}^*(j\omega) = 0$, the velocity is frequency independent and has the value c_{L_0} given by the expression

$$(c_{L_0})^2 \rho = K_0 = M_0 \qquad (6.9)$$

6. STRUCTURAL AND DIELECTRIC RELAXATION

For a purely viscous liquid, in which absorption is due only to the shear viscosity, equation (6.8) becomes

$$2\rho \, c_{L_0}^3 \, \alpha/\omega = \tfrac{4}{3} G''(\omega) = \tfrac{4}{3} \omega \eta$$

where η is the shear viscosity $(G''(\omega) \underset{\omega \to 0}{=} \omega\eta)$. Then

$$\alpha/\omega^2 = \frac{2\eta}{3\rho \, c_{L_0}^3} \qquad (6.10)$$

The quantity α/ω^2 (or α/f^2) is thus independent of frequency and represents the classical absorption due to shear viscosity. Absorption in excess of this can be attributed to a viscosity associated with compressional effects and called the volume viscosity, η_v. It is related to $K''(\omega)$ by the expression

$$\eta_v = \underset{\omega \to 0}{\text{Lim}} \, \frac{K''(\omega)}{\omega}$$

Then from equation (6.8), in the low frequency limit,

$$\underset{\omega \to 0}{\text{Lim}} \, 2\rho \, c_{L_0}^3 \alpha/\omega = \omega(\eta_v + \tfrac{4}{3}\eta)$$

or
$$\underset{\omega \to 0}{\text{Lim}} \, \alpha/\omega^2 = \frac{1}{2\rho \, c_{L_0}^3} (\eta_v + \tfrac{4}{3}\eta) \qquad (6.11)$$

The ratio of the observed to the classical absorption in the non-dispersion region enables the volume viscosity to be determined,

$$\frac{\alpha}{\alpha_{\text{class}}} = 1 + \frac{3}{4} \frac{\eta_v}{\eta} \qquad (6.12)$$

Variation of the value of α/f^2 with frequency indicates the presence of a relaxation process. In the simplest case, for a single relaxation process the bulk modulus is given by

$$K^*(j\omega) = K_0 + K_2 \frac{j\omega\tau_v}{1 + j\omega\tau_v} \qquad (6.13)$$

where τ_v is the volume relaxation time given by

$$\tau_v = \eta_v/K_2 \qquad (6.14)$$

The limiting value of $K^*(j\omega)$ at high frequencies is denoted by K_∞;

$$\lim_{\omega \to \infty} K^*(j\omega) = K_\infty = K_0 + K_2 \qquad (6.15)$$

The high frequency limiting value of the velocity is then given by

$$c_{L\infty}{}^2 \rho = M_\infty = K_0 + K_2 + \tfrac{4}{3} G_\infty. \qquad (6.16)$$

A distribution of relaxation times is normally required to describe adequately the observed bulk properties of a liquid. The variation of the complex bulk modulus with frequency is then described by an expression which parallels that used for the shear modulus,

$$K^*(j\omega) = K_0 + \int_0^\infty H_v(\tau) \frac{j\omega\tau}{1 + j\omega\tau} \, d\tau. \qquad (6.17)$$

$H_v(\tau)d\tau$ is then the contribution to the bulk modulus associated with volume relaxation times lying in the range τ to $\tau + d\tau$. It follows that

$$K_2 = \int_0^\infty H_v(\tau) d\tau.$$

As an alternative to the use of the bulk modulus $K^*(j\omega)$ is it possible to describe the liquid properties in terms of a complex compressibility $\beta^*(j\omega) = 1/K^*(j\omega)$. The equations relating $\beta^*(j\omega)$ to a spectrum of volume retardation times are similar in form to these for the complex shear compliance $J^*(j\omega)$ given in Chapter 3.

6.3 Analysis of Structural Relaxation Data

The experimental frequency range over which longitudinal wave absorption and velocity measurements may be made is normally insufficient to allow complete coverage of the whole relaxation region at any one temperature. Consequently, measurements are made over a range of temperatures and the data reduced to a single curve covering a wider range of effective frequency following the principles discussed in Chapter 3 for shear wave measurements.

A reduced value of the real part of the longitudinal modulus $\mathscr{M}'(\omega)$ may be defined

$$\mathscr{M}'(\omega) = \frac{M'(\omega) - M_0}{M_\infty - M_0} = \frac{K'(\omega) + \tfrac{4}{3} G'(\omega)}{K_2 + \tfrac{4}{3} G_\infty} \qquad (6.18)$$

where M_0 ($= \rho c_{L_0}^2$) is the static, or low frequency, modulus and M_∞ ($= \rho c_{L_\infty}^2$) is the high frequency limiting modulus. c_{L_0} and c_{L_∞} are the low and high-frequency limiting values of the velocity, respectively. The quantity $M_\infty - M_0 = M_2$ is the relaxational contribution to the total modulus.

The calculation of $\mathcal{M}'(\omega)$ thus requires that both M_∞ and M_0 be known as a function of temperature over the relaxation region. Velocity measurements made at low frequencies and high temperatures enable c_{L_0} to be measured over a range of temperature of 40K or more, from which the curve may be extrapolated to lower temperatures and into the relaxation region. Values of c_{L_0}, and its temperature dependence, have been tabulated by Nozdrev (1965) and Schaafs (1967). In many liquids the bulk modulus $K_0 (= \rho c_{L_0}^2)$ has been assumed to vary linearly with temperature, and the data extrapolated on this basis (Meister et al., 1960, Taşköprülü et al., 1961). Recently, however, precise measurements of c_{L_0} as a function of temperature (Yazgan, 1966; Barlow and Yazgan, 1966) have established that c_{L_0}, rather than K_0, is a linear function of temperature in many liquids. Extrapolation of c_{L_0} as a linear function of temperature has been found by Barlow et al. (1969c) to lead to a better reduction of their data at high effective frequencies, where the difference in the value of K_0 given by the two extrapolation procedures is the greatest.

The variation of c_{L_∞} with temperature can in principle be determined from velocity measurements made at low temperatures and high frequencies. Under these conditions the attenuation of the longitudinal wave is very high and the velocity cannot be determined accurately using the normal pulse technique. Consequently the variation of c_{L_∞} with temperature is often poorly defined, although the light scattering techniques described in Section 4.5 would appear to offer an alternative method of determining c_{L_∞}, often referred to in this context as the "hypersonic" velocity. Litovitz and Davies (1965, pp. 329–331) show that satisfactory reduction of the data is possible if both volume and shear relaxation processes have the same temperature dependence. That is, if the ratio of the mean relaxation times $\bar\tau_v/\bar\tau_s$ is independent of temperature, and if the ratio of the limiting moduli G_∞/M_2 is independent of temperature so that

$$1/G_\infty \left[\frac{\partial G_\infty}{\partial T}\right] = 1/M_2 \left[\frac{\partial M_2}{\partial T}\right] = 1/K_2 \left[\frac{\partial K_2}{\partial T}\right] \qquad (6.19)$$

If G_∞ is known as a function of temperature then equation (6.19) may be used to estimate the temperature variation of M_2. When the variation of M_2 with temperature cannot be determined directly, or from G_∞ via equation (6.19), an arbitrary variation of M_2 with temperature may have to be assumed, on the basis of obtaining the best reduction of the data to a single master curve. This method is less satisfactory, and the resulting reduced curve is prone to significant errors.

The ratio $\bar{\tau}_v/\bar{\tau}_s$ is equal to $\eta_v/\eta_s \times G_\infty/K_2$, so that if the above conditions for reduction are fulfilled the ratio η_v/η_s will also be independent of temperature. The values of the reduced variable $\mathscr{M}'(\omega)$ may thus be plotted against the variable $\omega\eta_s/G_\infty$ previously used as a normalised variable for plotting shear data.

The imaginary part of the longitudinal modulus $M'(\omega)$ may be plotted in reduced form by defining the variable

$$\mathscr{M}''(\omega) = \frac{M''(\omega)}{M_\infty - M_0} = \frac{K''(\omega) + \tfrac{4}{3}G''(\omega)}{K_2 + \tfrac{4}{3}G_\infty} \qquad (6.20)$$

The same requirements for reduction apply to $\mathscr{M}''(\omega)$ as to $\mathscr{M}'(\omega)$.

When the complex shear modulus $G^*(j\omega)$ is known, the complex bulk modulus $K^*(j\omega)$ can be determined by subtracting the effect of the shear process from the longitudinal data, as $M^*(j\omega) = K^*(j\omega) + \tfrac{4}{3}G^*(j\omega)$. The components $K'(\omega)$ and $K''(\omega)$ of the bulk modulus may then be plotted in reduced form as

$$\mathscr{K}'(\omega) = \frac{K'(\omega)}{K_2} \; ; \quad \mathscr{K}''(\omega) = \frac{K''(\omega)}{K_2} \qquad (6.21)$$

where $K_2 = K_\infty - K_0$. The values of K_2/G_∞ and η_v/η_s determined in this way to give satisfactory reduction of the data for $K'(\omega)$ may be tested by using the same values to plot the $K''(\omega)$ on a reduced plot. Satisfactory reduction of this independent data then lends support to the method of analysis and the assumptions made in the reduction process. By way of illustration, the variation of attenuation and velocity with temperature and frequency are shown in Figs. 6.1 to 6.7 for the idealised situation of single relaxation processes for both bulk and shear moduli. The curves have been computed from equations (6.3), (6.5), (6.13), (6.14) and (3.14) for the following values of the moduli and viscosities; the temperature variations used are typical of those found in supercooled liquids.

$K_2 = K_\infty - K_0 = \frac{4}{3} G_\infty$

$1/G_\infty = [1/1.6 + 0.01\ (T - 150)] \times 10^{-9}\ m^2 N^{-1}$

$\eta_v = 3\ \eta_s$

$\ln \eta_s = -11.7 + 1300/(T - 150)\ (\eta_s\ \text{in}\ N\ s\ m^{-2})$

$\rho = 10^3 \times (1 - 7 \times 10^{-4}(T - 150))\ kg\ m^{-3}$

$V_0 = 1900 \times (1 - 1.7 \times 10^{-3}(T - 150))\ m\ s^{-1}$

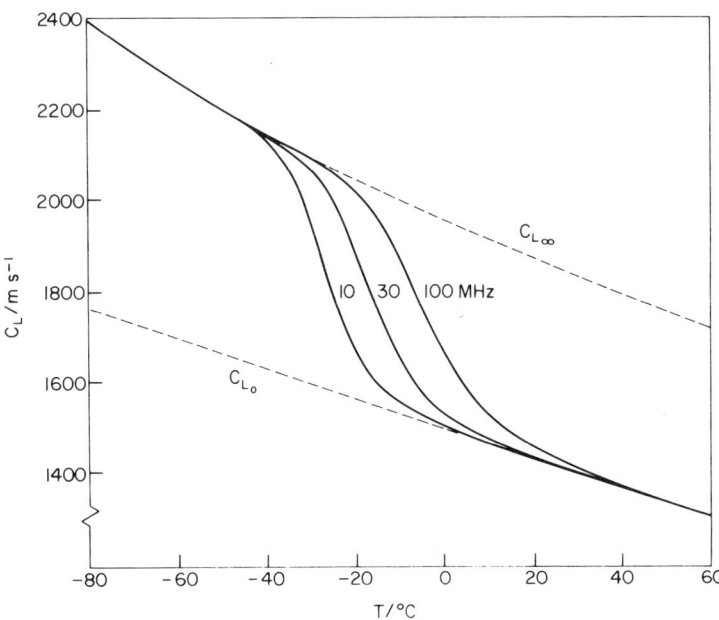

FIG. 6.1. Calculated variation of longitudinal wave velocity c_L as a function of temperature for a liquid having single structural and shear relaxation times.

Figures 6.3 to 6.7 illustrate the three common ways of presenting absorption data. The measured attenuation α, in units of nepers cm^{-1} or dB cm^{-1} (where 1 neper = 8.686 dB), can conveniently be plotted as a function of temperature at a fixed frequency, but the inherent dependence on the square of the frequency results in an inconveniently large range of values when data at different fre-

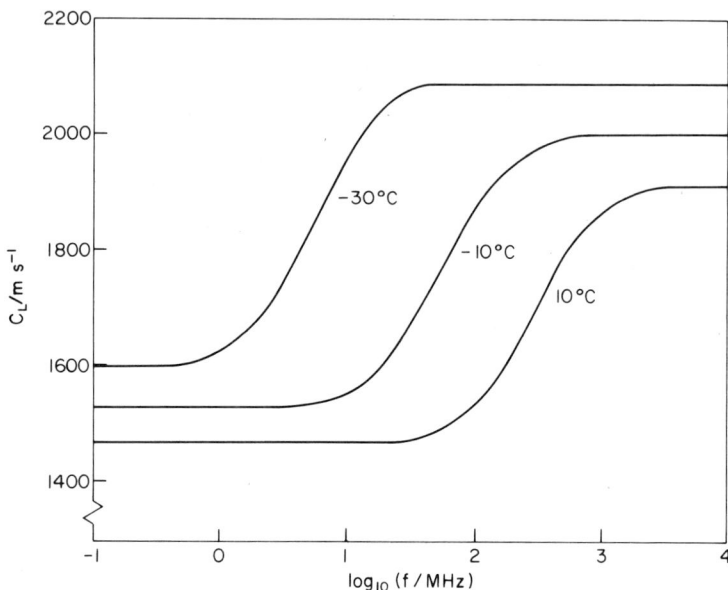

FIG. 6.2. Calculated variation of longitudinal wave velocity c_L as a function of frequency for a liquid having single structural and shear relaxation times.

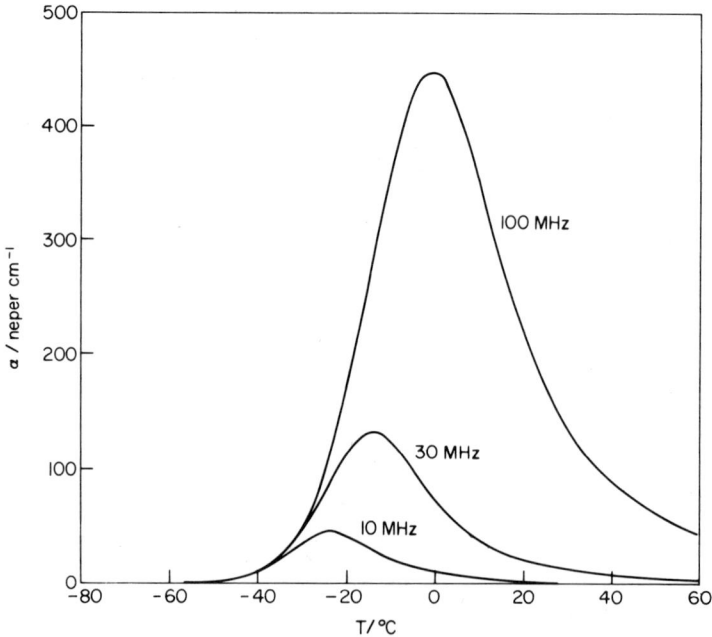

FIG. 6.3. Calculated variation of longitudinal wave attenuation α as a function of temperature for a liquid having single structural and shear relaxation times.

6. STRUCTURAL AND DIELECTRIC RELAXATION

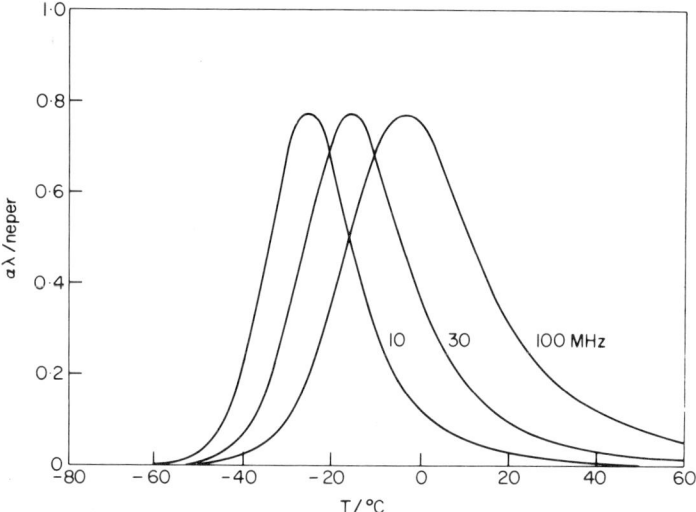

FIG. 6.4. Calculated variation of attenuation per wavelength $\alpha\lambda$ as a function of temperature for a liquid having single structural and shear relaxation times.

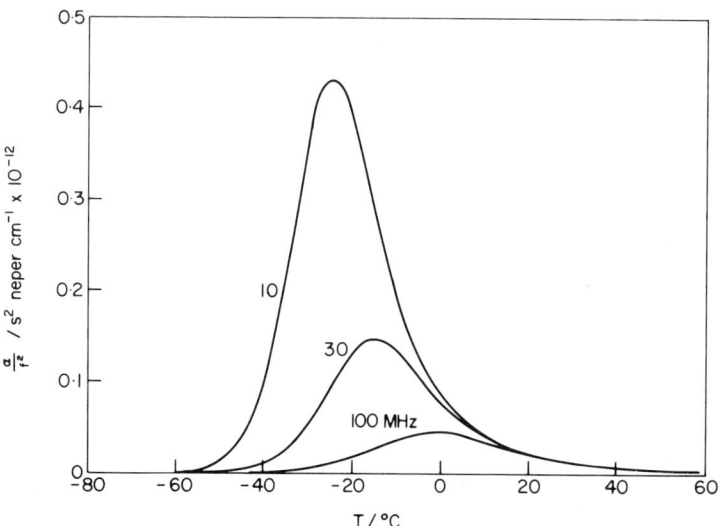

FIG. 6.5. Calculated variation of α/f^2 as a function of temperature for a liquid having single structural and shear relaxation times.

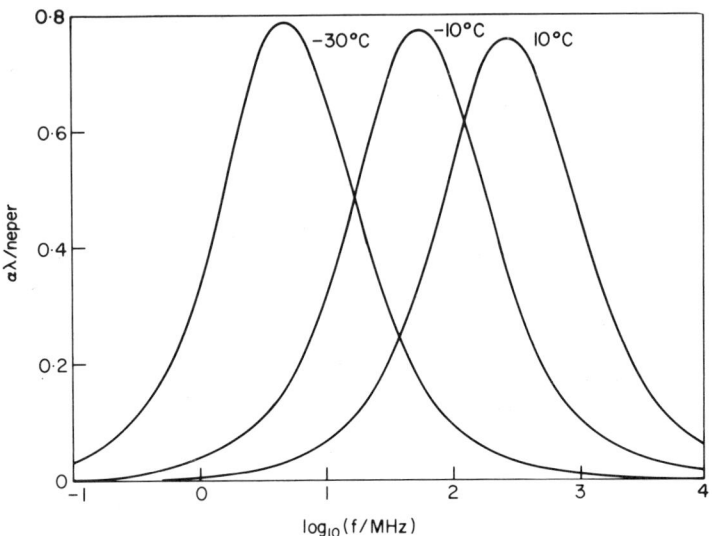

FIG. 6.6. Calculated variation of attenuation per wavelength $\alpha\lambda$ as a function of frequency for a liquid having single structural and shear relaxation times.

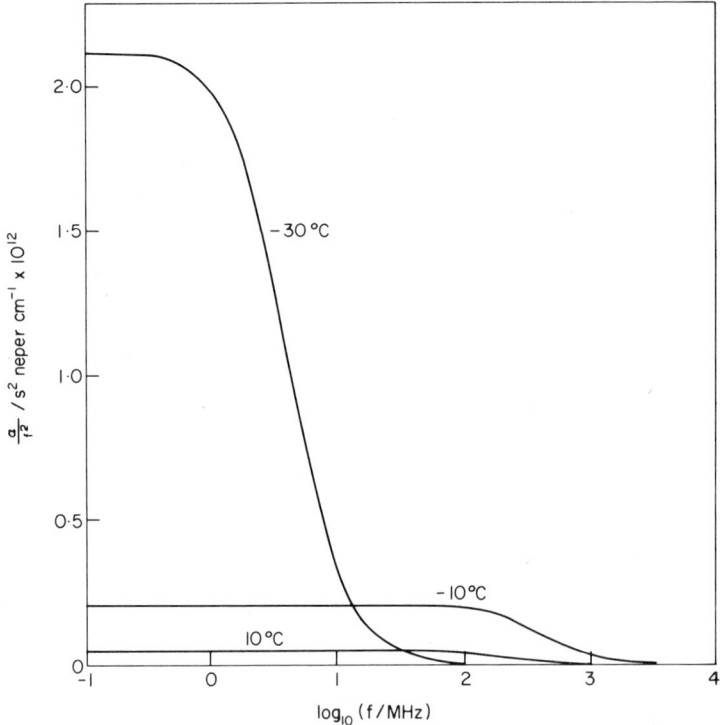

FIG. 6.7. Calculated variation of α/f^2 as a function of frequency for a liquid having single structural and shear relaxation times.

quencies are to be plotted on the same graph. The alternative quantities $\alpha\lambda$ (the absorption per wavelength) or α/f^2 remove this limitation, and are useful when data is plotted as a function of frequency. The quantity α/f^2 becomes independent of frequency at frequencies below the relaxation region and is then proportional to the combined effect of the shear and bulk viscosities (equation 6.11). The variation with frequency is similar to that of the dynamic viscosity η' in a shear relaxation process. The curves demonstrate clearly the ease with which the relaxation region can be explored with only a limited frequency range by making measurements over a range of temperatures.

The corresponding variations of the moduli with temperature and frequency are shown in Figs. 6.8 to 6.15. Figures 6.10, 6.11, 6.13 and 6.15 show the relative contribution of the shear and bulk processess to the total longitudinal modulus.

6.4 Structural Relaxation in Supercooled Liquids

The bulk, or volume, viscosity may be determined from measurements of the absorption coefficient α below the relaxation region. In this region the quantity α/f^2 is independent of frequency, and equation (6.11) can be used to determine η_v. The volume velocity is assumed to be associated with structural changes in the liquid, when molecular flow occurs between regions of high and low density produced by an elastic longitudinal wave. It is also assumed, in deriving equation (6.11), that the bulk and shear viscosities are the only factors contributing to the measured absorption α. There are, however, several other processes which result in a loss of energy from a longitudinal wave, with a consequent increase in the attenuation coefficient, and in these circumstances the estimation of the bulk viscosity can be difficult.

Allowance must be made for the absorption caused by the thermal conductivity of the liquid. The contribution to the absorption in a liquid of thermal conductivity K is

$$\alpha = \frac{\omega^2(\gamma - 1)K}{2\rho_0 c_{L_0}{}^3 C_p}$$

where C_p is the specific heat at constant pressure, and γ is the ratio of the specific heats (Herzfeld and Litovitz, 1959). Except in liquid metals, this contribution to the total absorption is usually only a few

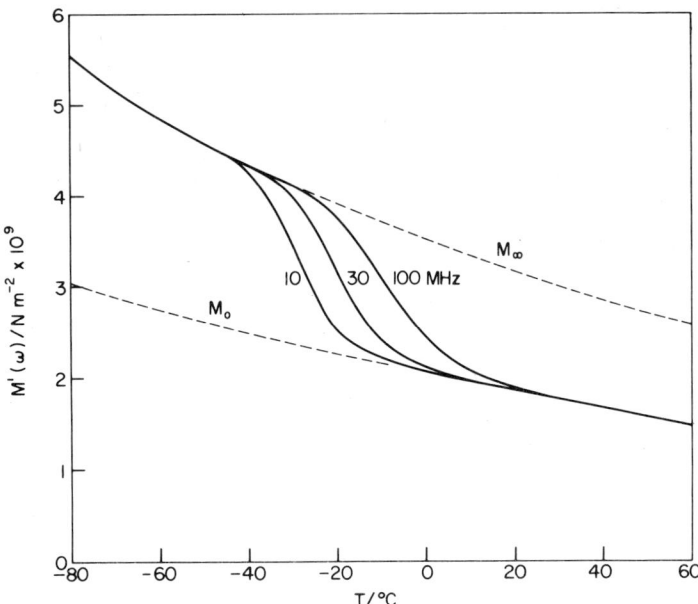

FIG. 6.8. Calculated variation of the real part $M'(\omega)$ of the longitudinal modulus $M^*(j\omega)$ as a function of temperature for a liquid having single structural and shear relaxation times.

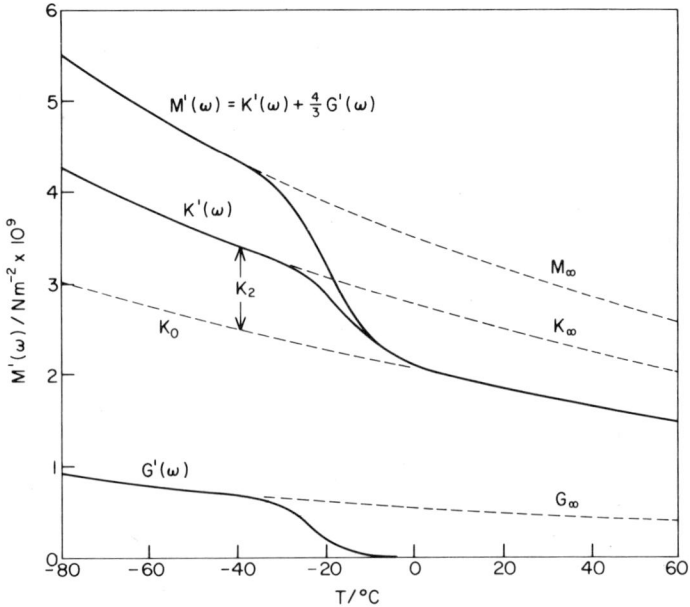

FIG. 6.9. Calculated variation of the real part $M'(\omega)$ of the longitudinal modulus $M^*(j\omega)$ as a function of temperature at a frequency of 30 MHz, showing the relative contributions of the shear, $G'(\omega)$, and structural, $K'(\omega)$, relaxation processes.

6. STRUCTURAL AND DIELECTRIC RELAXATION

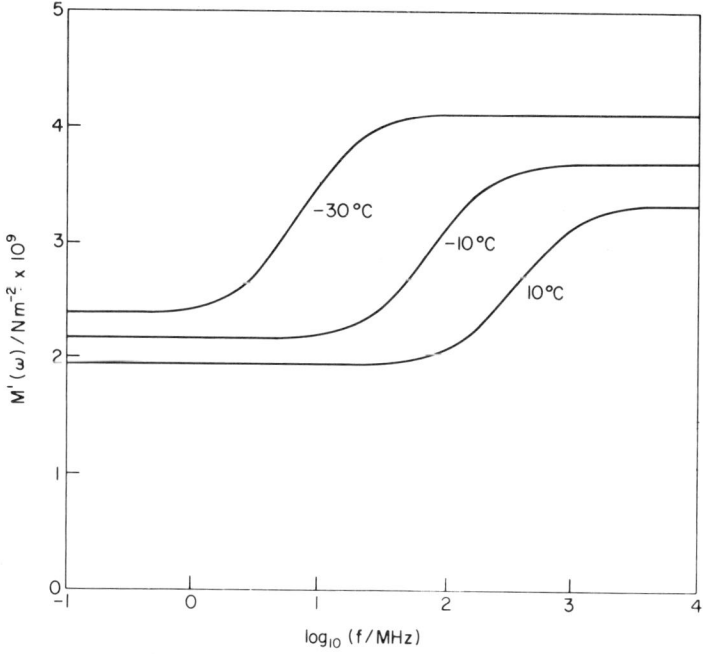

FIG. 6.10. Calculated variation of the real part $M'(\omega)$ of the longitudinal modulus $M^*(j\omega)$ as a function of frequency for a liquid having single structural and shear relaxation times.

percent. More important processes are these of rotational isomerism and chemical equilibria; a detailed discussion of these effects may be found elsewhere (Lamb, 1965; Matheson, 1971a; Herzfeld and Litovitz, 1959). At high temperatures the contribution of molecular vibration to the specific heat can lead to an increase in α/f^2 even though the frequency of measurement may be well below the vibrational relaxation frequency (Piercy and Rao, 1967). The relative magnitudes of these processes, and their different temperature dependences are well illustrated by the schematic diagram of Clark and Litovitz (1960) for isoamyl bromide, Fig. 6.16.

When allowance has been made for such additional contributions to the total absorption, the true "structural" viscosity is generally found to be between 1 and 3 times the shear viscosity, and to have a similar dependence on both temperature and pressure, (Litovitz and Davies, 1965; Piercy and Rao, 1967). The close similarities between the magnitudes of the shear and bulk viscosities and their temperature and pressure dependencies suggests that both viscosities have

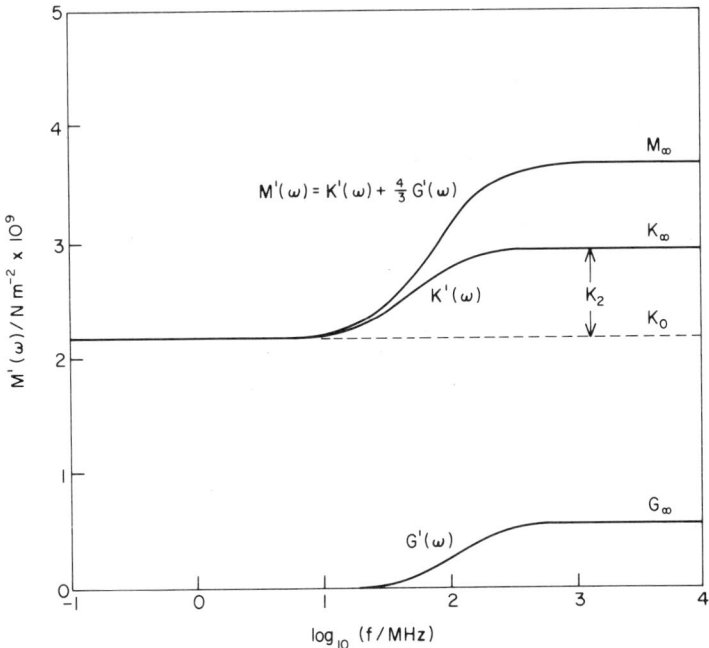

FIG. 6.11. Calculated variation of the real part $M'(\omega)$ of the longitudinal modulus $M^*(j\omega)$ as a function of frequency at a temperature of $-10°$C, showing the relative contributions of the shear, $G'(\omega)$, and structural $K'(\omega)$ relaxation processes.

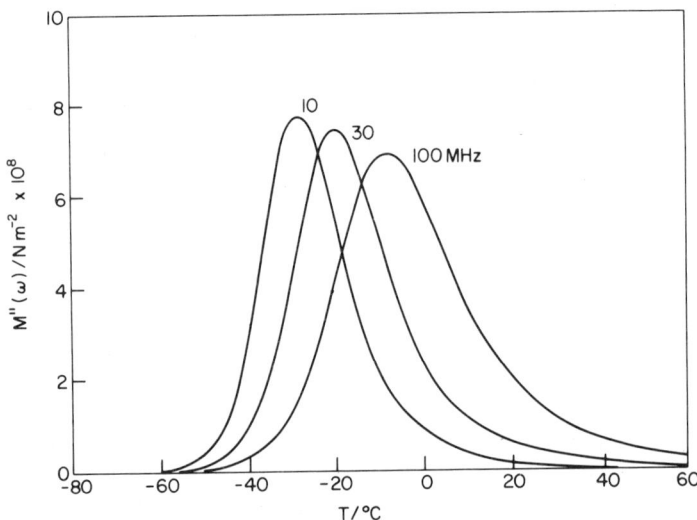

FIG. 6.12. Calculated variation of the imaginary part $M'(\omega)$ of the longitudinal modulus $M^*(j\omega)$ as a function of temperature for a liquid having single structural and shear relaxation times.

6. STRUCTURAL AND DIELECTRIC RELAXATION

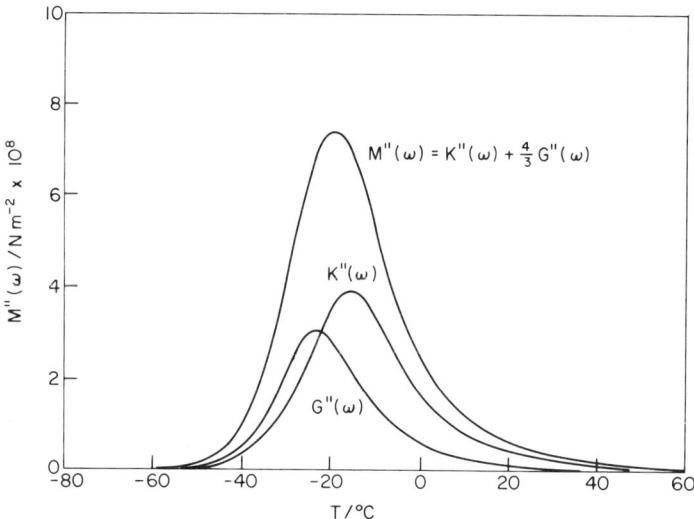

FIG. 6.13. Calculated variation of the imaginary part $M''(\omega)$ of the longitudinal modulus $M^*(j\omega)$ as a function of temperature at a frequency of 30 MHz, showing the relative contributions of the shear, $G''(\omega)$, and structural $K''(\omega)$ relaxation processes.

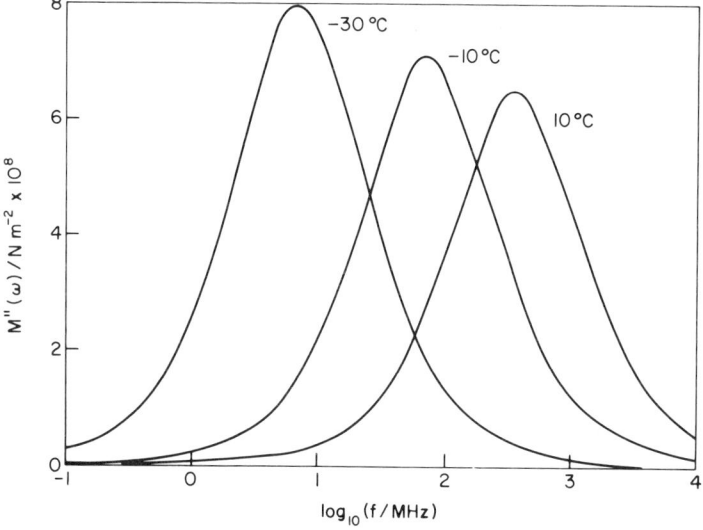

FIG. 6.14. Calculated variation of the imaginary part $M''(\omega)$ of the longitudinal modulus $M^*(j\omega)$ as a function of frequency for a liquid having single structural and shear relaxation times.

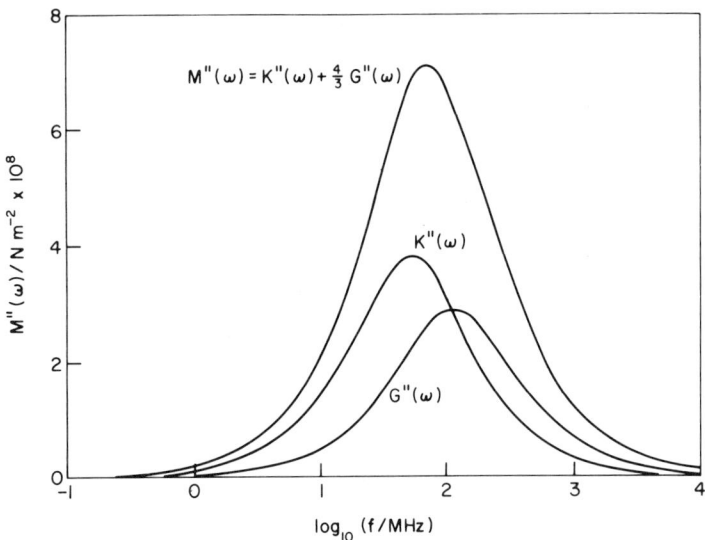

FIG. 6.15. Calculated variation of the imaginary part $M''(\omega)$ of the longitudinal modulus $M^*(\omega)$ as a function of frequency at a temperature of $-10°$C, showing the relative contributions of the shear, $G''(\omega)$, and structural, $K''(\omega)$, relaxation processes.

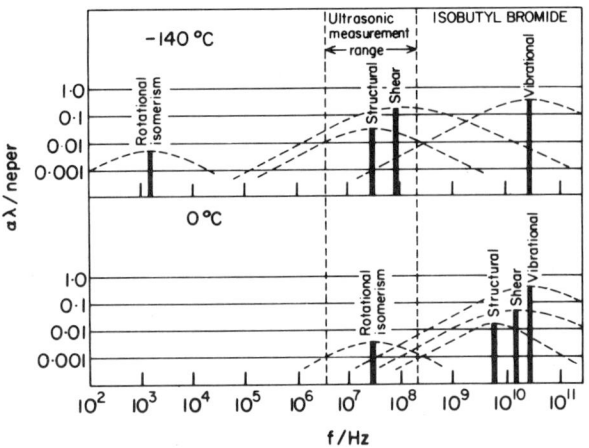

FIG. 6.16. Spectrograms of the loss per wavelength, $\partial\lambda$, in isobutyl bromide due to rotational isomerism, vibrational, shear and structural mechanisms at temperatures of $-140°$C and $0°$C (after Clark and Litovitz, 1960).

their origins in the hindered motion of the molecules. Measurements of the structural relaxation process in supercooled liquids show that this is very similar to the shear relaxation process. In two liquids, bis[m-(m-phenoxy phenoxy)phenyl] ether (Barlow et al., 1969c) and tri-tolyl phosphate (Barlow and Singh, 1972), both the shear and structural relaxations have the same form and can be described by the empirical equation of Barlow et al. (1967b). The modulus $K^*(j\omega)$ is given by an equation similar in form to equation (5.3).

$$\frac{K^*(j\omega) - K_0}{K_2} = [1 + K_2/j\omega\eta_v + 2(K_2/j\omega\eta_v)^{1/2}]^{-1} \quad (6.22)$$

Figures (6.17) and (6.18) show the normalised values of bulk modulus $K'(\omega)/K_2$ and loss modulus $M''(\omega)/K_2$, plotted against normalised frequency $\omega\eta/G_\infty$ for bis[m-(m-phenoxy phenoxy)phenyl] ether.

These results differ in detail from the earlier measurements of Litovitz and co-workers on a range of associated liquids. These

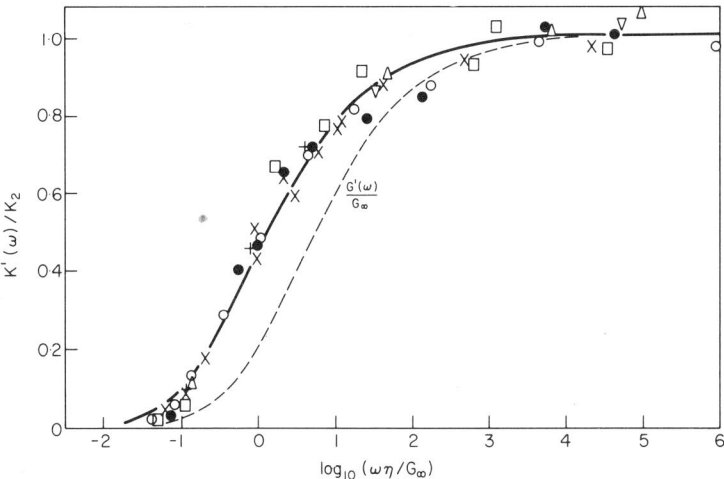

FIG. 6.17. Values of the normalised bulk modulus $K'(\omega)/K_2$ plotted as a function of $\log_{10}(\omega\eta/G_\infty)$ for bis(m-(m-phenoxy phenoxy)phenyl)ether. The dashed line shows the variation of the normalised shear modulus $G'(\omega)/G_\infty$ calculated using equation (5.3). The solid line shows the variation of $K'(\omega)/K_2$ calculated using equation (6.22) assuming that $\eta_v = 4 \times \frac{4}{3}\eta$ and $K_2 = \frac{4}{3}G_\infty$ (after Barlow et al. 1969c).

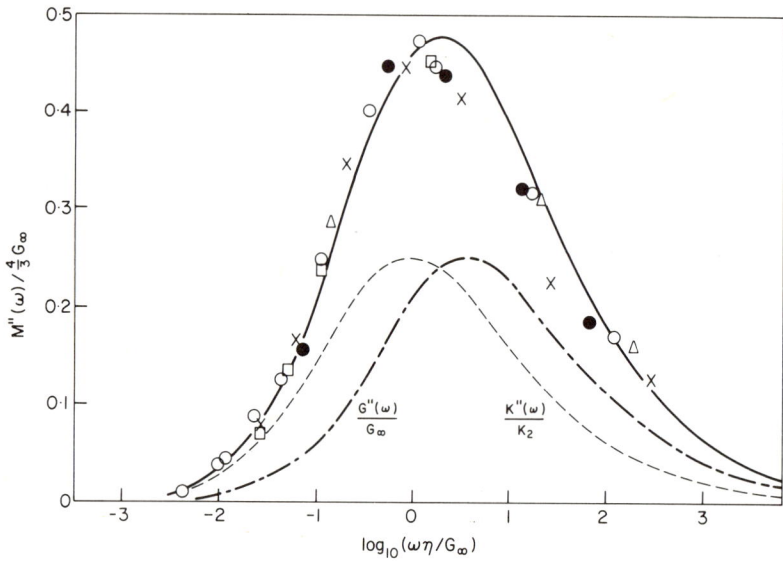

FIG. 6.18. Normalised values of the longitudinal loss modulus $M''(\omega) = K''(\omega) + \frac{4}{3}G''(\omega)$ plotted as a function of $\log_{10}(\omega\eta/G_\infty)$ for bis(m-(m-phenoxy phenoxy)phenyl)ether. The solid line shows the variation of $M''(\omega)$ calculated using equations (5.3) and (6.22) assuming that $\eta_v = 4 \times \frac{4}{3}\eta$ and $K_2 = \frac{4}{3}G_\infty$. The dashed curves show the individual contributions of $K''(\omega)$ and $G''(\omega)$ to the total normalised modulus (after Barlow et al. 1969c).

workers find that the shapes of the shear and compressional relaxation curves are different, and that different distribution functions of relaxation times are required to describe the experimental data (Meister et al. 1960). These differences are surprising in view of the apparently close connection between the two relaxation processes. Furthermore, discrepancies are found between the measured values of $M''(\omega)$ and the values calculated from the distribution functions which are used to describe the $K'(\omega)$ and $G'(\omega)$ data. As shown in Fig. 6.19, at frequencies above the peak in the loss curve the measured values lie consistently above the calculated curve, the difference tending towards a constant value at high frequencies. Litovitz and Lyon (1954), observing this type of behaviour in glycerol and bi-phenyl pentachloride, attributed this extra loss to a "hysteresis" mechanism which gives a constant loss per cycle, independent of frequency and viscosity. No molecular basis for this mechanism has been proposed.

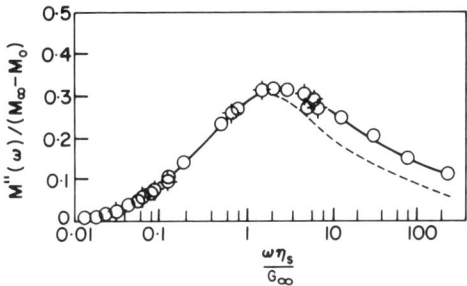

FIG. 6.19. Measured variation of $M''(\omega)$ as a function of $\log_{10}(\omega\eta/G_\infty)$ for 1,3-butanediol. The dashed line is calculated from the distribution functions of shear and structural relaxation times used to describe the measured values of $M'(\omega)$ (after Meister et al. 1960).

Barlow et al. (1969c) do not find this additional high frequency loss in their measurements. In a detailed discussion, they suggest that the origin of the discrepancies may lie in the difficulties of reducing data obtained over a restricted temperature range. Errors in the extrapolations used for K_0, K_2 and G_∞ can lead to significant distortion in the shape of the reduced curve, especially in the high frequency region of the curve.

These difficulties, already discussed for shear relaxation data in Section 5.2, assume a greater significance for longitudinal data. Not only is the relaxing part (K_2) of the bulk modulus a small part of the total bulk modulus $K'(\omega)$, and this in turn only a part of the observed longitudinal modulus $M'(\omega)$, (see Fig. 6.9 and 6.10), but the possibility of additional processes such as rotational isomerism may further complicate the analysis (e.g. Dexter and Matheson, 1971). Matheson (1971a) points out that these processes do not necessarily contribute in a simple additive manner to the total absorption. Although simple additivity is commonly assumed, Truesdell (1953) has shown that this is only an approximation.

Great care must therefore be taken in analysing longitudinal relaxation measurements, and in drawing conclusions from the reduced curves. In particular, the assumption that the distributions of relaxation times remain independent of temperature and pressure is likely to be true only over limited ranges only. Conventional measurements made over a large temperature range on the molten salt boron trioxide show that the width of the distribution of relaxation times decreases with increasing temperature (Tauke *et al.*, 1968). Similar behaviour is found in glycerol, using Brillouin scattering to enable measurements within the relaxation range to be obtained at higher temperatures than is possible using the conventional pulse technique (Pinnow *et al.*, 1968a). These trends are similar to the temperature dependences observed in viscoelastic and dielectric relaxation, but the data are insufficient to enable the changes in the width of the spectrum to be specified accurately.

Further measurements are clearly required to determine whether the similarity between the shear and structural relaxation curves is a general property of supercooled liquids. It has been shown, in Section 5.2 that equation (5.3) is only an approximate description of the relaxation process, and more correct representation is in terms of retardation parameters, equation (5.11). In view of the similarity of the shear and structural properties it may be inferred that the compressional modulus should be represented by the equation

$$\frac{K^*(j\omega) - K_0}{K_2} = \frac{1}{1 + \dfrac{K_2}{j\omega\eta_v} + \dfrac{J_{r_v} K_2}{(1 + j\omega\tau_{r_v})^\beta}} \quad (6.23)$$

where J_{r_v} is a retardational compliance and τ_{r_v} a retardation time. Because of the difficulties discussed above, it is unlikely that measurements of sufficient accuracy can be obtained to distinguish between equations (6.22) and (6.23), and to determine J_{r_v} and τ_{r_v} in equation (6.23).

Nevertheless, a knowledge of K_0, K_2 and η_v, and their variation with temperature and pressure, is of considerable practical significance. The lubricant between two highly loaded surfaces is subjected to pressures of the order of 1 GN m^{-2} for periods as short as 10^{-4} s. The time required for the liquid to reach a new equilibrium state after a pressure change is governed by the characteristic time $\tau_v = \eta_v / K_2$ where η_v and K_2 are the values at the maximum pressure. The transit time of the lubricant between the surfaces may often be less than the time required to reach equilibrium, and the shear viscosity will not then attain the equilibrium value corresponding to the pressure experienced. Harrison and Trachman (1972) have shown, using a simple model with a single but pressure dependent relaxation time, that by incorporating the effects of structural relaxation into an analysis of elasto-hydrodynamic lubrication, better agreement can be obtained between the theoretical and the observed behaviour in rolling contact systems. In particular, this type of analysis provides an explanation for the fall in the effective viscosity of a lubricant between two heavily loaded rollers when the rolling speed is increased. This effect has been observed by several investigators (Crook, 1963; Bell et al., 1964; Johnson and Cameron, 1967) and is present when temperature and shear rate effects are negligible. There is thus considerable interest in determining the basic structural relaxation parameters K_0, K_2 and η_ν at pressures of the order of 1 GN m^{-2}, even though it may not be possible to define the relaxation process in detail.

6.5 The Dielectric Properties of Supercooled Liquids

The phenomenon of dielectric relaxation has been studied, as a means of investigating molecular motions, ever since the pioneering work of Debye (1913). The literature is consequently extensive and both the general theory and details of the experimental techniques, as well as the results of dielectric measurements on gases, liquids and solids, are available in review articles and books (see for example: Smyth, 1955; Hill et al., 1969). The present discussion is confined to the dielectric behaviour of supercooled liquids and to the possible correlations between the dielectric and the mechanical properties of liquids.

A polar liquid is one containing molecules having permanent dipole moments, which in the absence of an electric field are distributed randomly in all directions. When an electric field is applied, there is a tendency for the molecules to rotate so that the dipoles become aligned parallel to the field, resulting in an increase in the permittivity ϵ. The thermal motion of the molecules allows only a small departure from the random configuration and the permittivity falls with increasing temperature. The value of the permittivity under a steady field, at a given temperature, is denoted by ϵ_0. If an alternating field is applied the orientation of the molecules will be able to follow the reversals of the field if the frequency is sufficiently low and the permittivity will remain constant at the value ϵ_0. As the frequency is raised, the dipoles no longer have time to reach their equilibrium position and the permittivity falls, reaching a limiting high frequency value ϵ_∞. This residual permittivity is associated with the electronic and atomic polarization which occurs in non-polar materials. The value remains constant until the applied field has a frequency approaching the natural frequencies of vibration of the atoms, which is usually in the infra-red region of the spectrum.

If all the molecules in a liquid respond with a characteristic time τ_D to the application of an electric field, then the response to an alternating field of angular frequency ω can be represented by a complex permittivity $\epsilon^*(j\omega) = \epsilon'(\omega) - j\epsilon''(\omega)$ where

$$\epsilon^*(j\omega) = \epsilon_\infty + \frac{\epsilon_0 - \epsilon_\infty}{1 + j\omega\tau_D}. \tag{6.24}$$

The components $\epsilon'(\omega)$ and $\epsilon''(\omega)$ of the complex permittivity are then given by

$$\epsilon'(\omega) = \epsilon_\infty + \frac{\epsilon_0 - \epsilon_\infty}{1 + \omega^2 \tau_D^2} \tag{6.25}$$

$$\epsilon''(\omega) = \frac{\epsilon_0 - \epsilon_\infty}{1 + \omega^2 \tau_D^2} \omega\tau_D. \tag{6.26}$$

The frequency dependence of the components, shown in Fig. 6.20, is thus identical to the frequency dependence of the components of the complex compliance $J^*(j\omega)$ for a process characterised by a single retardation time τ_r (equations 3.19, 3.20, 3.23). It follows that in comparing dielectric and mechanical data, the frequency dependence of $\epsilon^*(j\omega)$ should be compared with that of the complex compliance $J^*(j\omega)$ rather than the modulus $G^*(j\omega)$. Also, the process commonly termed dielectric "relaxation" should, by analogy with mechanical behaviour, be called dielectric "retardation".

It is an established convention that dielectric data is presented in parametric form. Equations (6.25) and (6.26) may be written

$$\frac{\epsilon'(\omega) - \epsilon_\infty}{\epsilon_0 - \epsilon_\infty} = \frac{1}{1 + \omega^2 \tau_D^2} \; ; \quad \frac{\epsilon''(\omega)}{\epsilon_0 - \epsilon_\infty} = \frac{\omega\tau_D}{1 + \omega^2 \tau_D^2}$$

when they form the parametric equations of a circle in the $\epsilon'(\omega)$, $\epsilon''(\omega)$ plane; the centre is at $\{(\epsilon_0 + \epsilon_\infty)/2), 0\}$ and radius is

6. STRUCTURAL AND DIELECTRIC RELAXATION

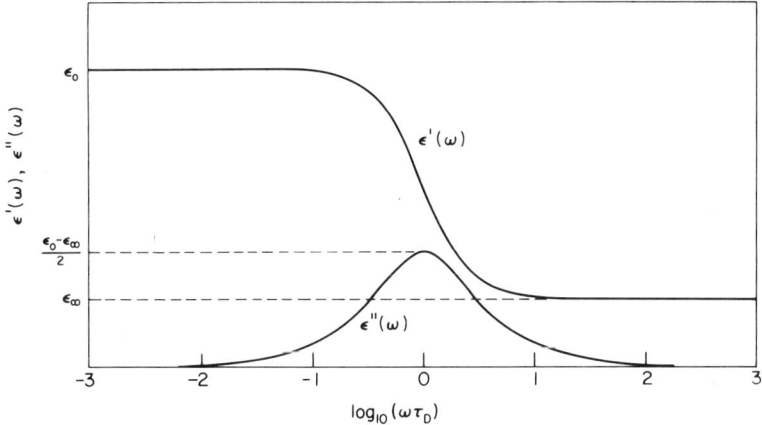

FIG. 6.20. Variation of the components of the complex dielectric permittivity $\epsilon^*(j\omega) = \epsilon'(\omega) - j\epsilon''(\omega)$ as a function of frequency for a relaxtion process having a single relaxation time τ_D.

$(\epsilon_0 - \epsilon_\infty)/2$. Results are normally plotted on the semi-circle lying above the $\epsilon'(\omega)$ axis, as shown in Fig. 6.21.

For many liquids of low viscosity the experimental data is found to lie on a semi-circle, and the relaxation process can then be characterised by a single time τ_D. (Buckley and Maryott, 1958). However, many viscous liquids, especially these which can be supercooled to near the glassy state, show a rather broader loss curve, with a maximum loss which is lower than the value $(\epsilon_0 - \epsilon_\infty)/2$

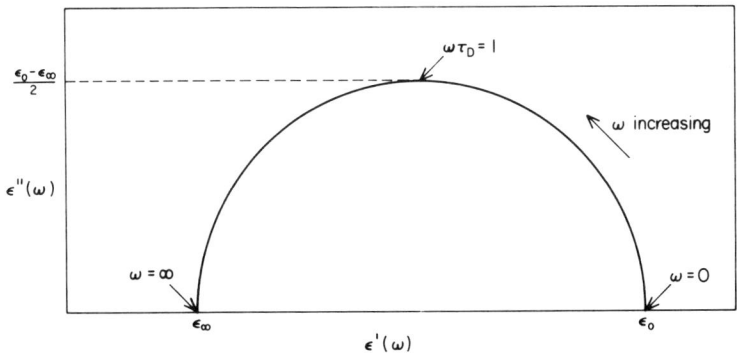

FIG. 6.21. Locus of the complex dielectric permittivity $\epsilon^*(j\omega)$ for a single relaxation time process — the Debye semicircle.

predicted by equation (6.26). An empirical equation to describe this form of behaviour was proposed by Cole and Cole (1949),

$$\epsilon^*(j\omega) = \epsilon_\infty + \frac{\epsilon_0 - \epsilon_\infty}{1 + (j\omega\tau_D)^\alpha} \qquad (6.27)$$

where α is a constant, $0 \leqslant \alpha \leqslant 1$. The curve in the ϵ', ϵ'' plane is a depressed arc with its centre below the ϵ' axis, the arc being symmetrical about a line through the point $\{(\epsilon_0 + \epsilon_\infty)/2, 0\}$ parallel to the ϵ'' axis. The quantity τ_D must now be regarded as a "mean", or "effective" relaxation time. Although the behaviour of many liquids can be described by equation (6.27), Davidson and Cole (1951) found that for glycerol and propylene glycol at low temperatures (c. $-60°$C) the experimental results did not display the symmetry of the Cole–Cole arc, but lay on a skewed arc when plotted on the ϵ', ϵ'' plane. The arc may be described by the equation

$$\epsilon^*(j\omega) = \epsilon_\infty + \frac{\epsilon_0 - \epsilon_\infty}{(1 + j\omega\tau_D)^\beta} \qquad (6.28)$$

where β is a constant between 0 and 1. Figure 6.22 shows the predictions of this equation for several values of β, the curve tending

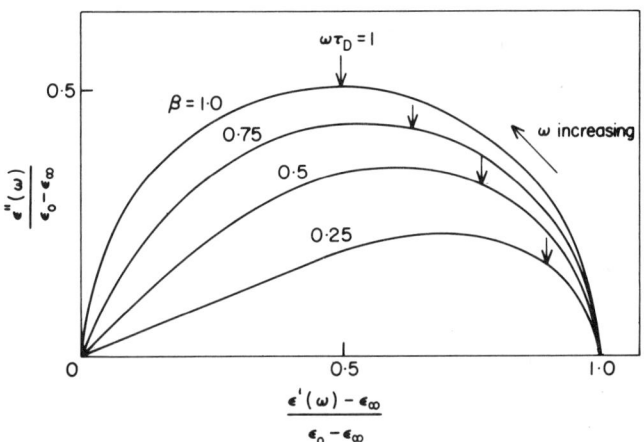

FIG. 6.22. Values of the normalised complex dielectric permittivity $(\epsilon^*(j\omega) - \epsilon_\infty)/(\epsilon_0 - \epsilon_\infty)$ calculated from the empirical equation of Davidson and Cole (1951) for several values of the parameter β (Equation 6.28).

towards the Debye semicircle as β approaches 1. Davidson (1961) has summarised the evidence in favour of this skewed-arc distribution, which provides a good description of the behaviour of many liquids, including glycerol, a range of other hydroxylic alcohols, halogenated alkane derivatives and polychlorinated biphenyl derivatives. A detailed and careful discussion of the adequacy of the skewed-arc plot is also given by Mopsik and Cole (1966) for the case of n-octyl iodide and other liquid alkyl halides.

The skewed-arc function may therefore be regarded as the general form for the dielectric behaviour of liquids, with the Debye semicircle as a limiting case when $\beta = 1$. The spectrum of times associated with this function has been discussed previously in connection with the viscoelastic behaviour of liquids, the form of the distribution is given by equation (5.12) and is shown in Fig. 5.13. Two different approaches have been used in attempting to account, in terms of molecular motions, for the presence of a distribution of times. Firstly, it may be assumed that reorientation of the dipole can take place by a variety of molecular motions, each characterised by a single time. Also, interactions between neighbouring molecules will result in differing local environments leading to further variations in the time scales of the motions. The result is a distribution of times, with the maximum time corresponding to reorientation of the molecule under conditions of maximum hindrance. This approach has been used by Higasi et al. (1960). Vaughan et al. (1962) and Matsumoto and Higasi (1962) in describing the behaviour of n-alkyl bromides, where a variety of modes for reorienting the dipole exist. The spectrum of relaxation times used is of the form $L(\tau) = 1/A\tau^n$ over the range $\tau_1 \leqslant \tau \leqslant \tau_2$. The shortest time τ_1 is associated with the rotation of the $CH_2 Br$ end-groups, and the longest time τ_2 with the slowest motion, the end-over-end rotation of the molecule. The exponent n is chosen to give the best fit to the experimental data and is typically in the range 0.5 to 1.0, when the plot of the complex permittivity is very similar to the Davidson—Cole skewed arc. At low temperatures the ratio τ_2/τ_1 increases, as τ_2 increases more rapidly with decreasing temperature than τ_1 as the glass transition is approached. The value of τ_1 always remains reasonably large, even at elevated temperatures, and remains consistent with the time of rotation of the end group in a viscous medium. The termination of the distribution at the shortest time τ_1 avoids the difficulty of providing a physical interpretation for the unterminated spectrum of the Davidson–Cole function at short times.

The second approach places the emphasis on the co-operative interactions between neighbouring molecules rather than on single-molecule behaviour. The model proposed by Glarum (1960b) involves the concept of a defect in the liquid structure, moving by a process of diffusion through the liquid. In the presence of such a defect, instantaneous reorientation of a molecule is possible, whereas in the absence of a defect reorientation can occur with a characteristic time τ'. The diffusion process is characterized by a diffusion time τ_d, and a range of different types of behaviour is predicted as the ratio τ'/τ_d is varied. The expression for the complex permittivity is given by

$$\frac{\epsilon^*(j\omega) - \epsilon_\infty}{\epsilon_0 - \epsilon_\infty} = \frac{1}{1 + j\omega\tau'} \left\{ 1 + \frac{j\omega\tau'}{1 + \sqrt{(1 + j\omega\tau')\tau_d/\tau'}} \right\} \quad (6.30)$$

If $\tau' \ll \tau_d$, the probability of a defect arriving in time τ' will be small, and a simple Debye process is predicted, with a single time constant τ'. Conversely, if $\tau_d \ll \tau'$, defect diffusion will be the dominant process, and equation (6.30) reduces to the equation of the Cole–Cole depressed circular arc. Between these two extremes, when the two times are comparable, the response is similar to the Davidson–Cole skewed-arc with the parameter β close to the value 0.5. This model implies that the diffusion of defects plays an important role in determining the rate at which molecular reorientation occurs in supercooled liquids, and does not rely on the presence of intra-molecular motions, although such motions may also be present. The Glarum model is based on a one-dimensional diffusion process and is thus somewhat unrealistic, although an extension to three dimensions leads to similar results (Hunt and Powles, 1966).

The emphasis in Glarum's model on the co-operative nature of the interactions between neighbouring molecules is supported by the behaviour of solid hydrogen iodide and bromide. The measurements of Groenewegen and Cole (1967) show that in these materials, in the crystalline phase, the dielectric behaviour follows the skewed-arc pattern. In this case the concept of a range of intra-molecular motions cannot be applied. Further evidence is provided by the behaviour of dilute solutions of small polar molecules in the almost non-polar supercooled solvent orthoterphenyl (Williams and Hains, 1973). For the six polar materials studied, the curves of $\epsilon''(\omega)$ plotted against frequency were found to be identical in shape. These loss curves were far broader than for a single relaxation time process and the shape was independent of concentration. The frequency of maximum loss, when extrapolated to zero concentration, and its temperature dependence, coincided with those of the mildly polar solvent. Williams and Hains conclude that the motion of the solute molecules reflects the motions of neighbouring solvent molecules moving in a co-operative manner, and that a similar process occurs in many other liquids. The frequency of maximum loss is generally found to vary rapidly with temperature, in a manner similar to the shear viscosity as described by the free volume equation (2.8) (Denney, 1957, Winslow et al., 1957; Johari and Goldstein, 1970). The loss curves measured by Williams and Hains can be described in terms of an empirical decay function

$$\phi(t) = \exp - \left(\frac{t}{\tau_0}\right)^{\bar{\beta}} \qquad (6.31)$$

where $0 \leq \bar{\beta} \leq 1$. (Williams and Watts, 1970). This function results in a frequency dependent complex permittivity which closely follows the Davidson–Cole skewed-arc form, although significant differences occur for low values of the parameter $\bar{\beta}$. In this case, and for the Glarum defect-diffusion model, no physical interpretation is attached to the distributed spectrum of relaxation times, which is regarded only as a formal mathematical representation of the process.

The width of the spectrum is generally found to decrease as the temperature is raised, tending towards a single relaxation time process at high temperatures. This change is consistent with the reduced interaction between molecules at temperatures well above the glass transition, when the viscosity is low and molecular motions are not free-volume controlled. In the supercooled region the change in width with temperature is often small, however, the value of the parameter β in the Davidson–Cole expression typically increasing from 0.50 to 0.60 in a 10K interval (Denney, 1957). The mean relaxation time changes rapidly in the supercooled region, increasing by up to three orders of magnitude for a 10K reduction in temperature. The similar behaviour of the viscosity with temperature suggests that these two properties may have a common origin in the local structure of the liquid.

Debye attempted to relate the dielectric relaxation time to the viscosity by assuming the molecule to be a sphere rotating in a continuous viscous medium having the same viscosity (η) as the bulk liquid. The relaxation time of an individual molecule is then

$$\tau' = \frac{3V\eta}{kT} \qquad (6.32)$$

where V is the molecular volume. The observed macroscopic time τ differs from the microscopic time τ' because of the influence on each molecule of the electrical field from neighbouring molecules. A variety of expressions relating the two times have been proposed, using different expressions for the internal field, but that proposed by Powles (1953) and also by Glarum (1960a) is commonly used. The microscopic time τ' is given in this case by

$$\tau' = \left[\frac{2\epsilon_0 + \epsilon_\infty}{3\epsilon_0}\right] \tau. \qquad (6.33)$$

The ratio τ'/τ thus has extreme values of 0.67 and 1.0. In many cases any uncertainty in this factor can be ignored when compared with the other uncertainties involved in the estimation of the macroscopic time τ.

Debye's original postulate, that the viscosity η in equation (6.32) is the same as the shear viscosity, can give the correct order of magnitude for τ, but usually over-estimates the value by up to a factor of 10. Measurements made on solutions and mixtures show the importance of the shape and size of a molecule relative to its neighbours. (Cole, 1960; Smyth, 1966). In many cases, a correlation is observed between the temperature dependence of the dielectric relaxation time and that of the viscosity. Litovitz and McDuffie (1963) have compared the dielectric relaxation times with the shear and compressional relaxation times in a range of associated liquids. Over a temperature range of 30K both the dielectric relaxation time τ_D and the ratio η/T changed by about two orders of magnitude. The ratio $\tau_D T/\eta$ changed significantly over this range, by up to 300%, and a much closer correlation was observed between τ_D and the average shear relaxation time. A similar close correlation was found between the dielectric and mechanical relaxation times over a pressure range of 55,000 psi. The similarity in the temperature dependences of the shear and volume relaxation processes has been discussed in the previous Section (6.4). Litovitz and McDuffie suggest that the similar dependences of the dielectric and the mechanical relaxation times on temperature and pressure indicate that these relaxation processes all have a common origin in the cooperative nature of molecular motions which occur in the supercooled state. A more detailed comparison of the dielectric and the mechanical processes is prohibited, however, by the method used to analyse the mechanical data. Litovitz and McDuffie draw an analogy between the complex permittivity and the complex shear and bulk moduli. However, as discussed previously (on page 174) the dielectric process is a *retardation* process (in mechanical terminology) and the analogous mechanical quantity to the complex permittivity is the complex *compliance*. Thus the dielectric "relaxation" time should be compared with the mechanical retardation time τ_r, and the width of the spectrum of times, expressed as a Davidson–Cole function, compared with the mechanical retardation spectrum rather than the relaxation spectrum.

The analysis of mechanical data in terms of a retardation spectrum is made difficult by the limited accuracy of the experimental results, and measurements of sufficient accuracy have been made on only a few liquids (see Chapter 5, Section 5.3). Shears et al. (1974) have compared the dielectric and viscoelastic behaviour of three liquids, tri(o-tolyl)phosphate, di(n-butyl)phthalate and tri(β-chloroethyl)-phosphate, and the data are shown in Fig. 6.23. The general forms of the locus plots for the dielectric and viscoelastic data are very similar. The largest discrepancy occurs with tri-(β-chloroethyl)phosphate, but the dielectric loss curves for this liquid show evidence of bimodal behaviour which is not present in the viscoelastic results. The agreement for the other two liquids is within experimental error, and there is also good agreement between the location of the maxima of the dielectric and viscoelastic loss curves as a function of frequency and temperature. This close agreement between the shape and the location of the dielectric and viscoelastic retardation process leads Shears *et al.* (1974) to conclude that as the dielectric behaviour is associated with the reorientation of polar molecules, the retardational part of the viscoelastic process must likewise be associated

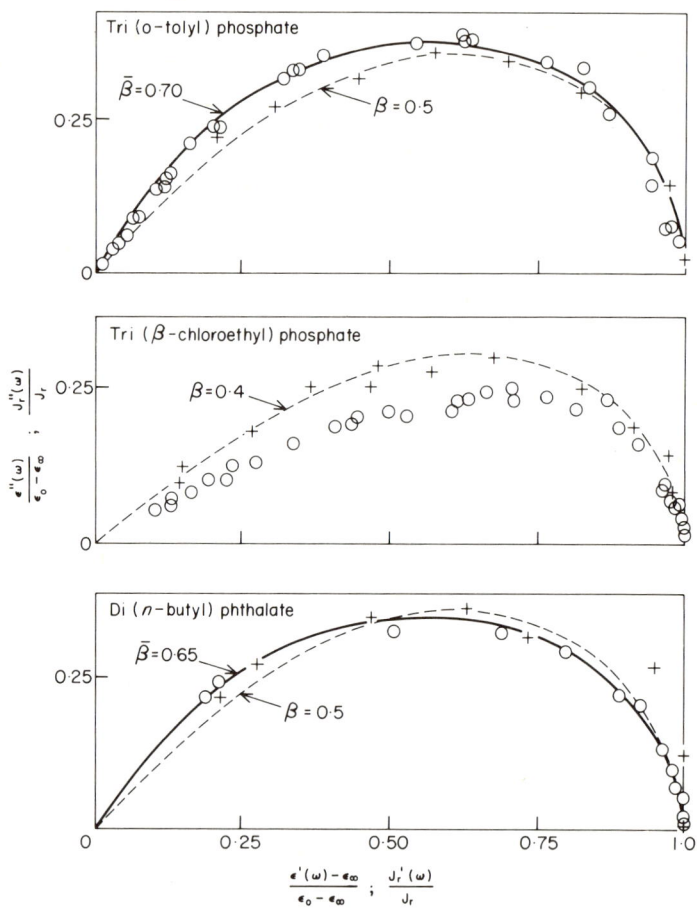

FIG. 6.23. Dielectric relaxation and viscoelastic retardation in tri(o-tolyl)phosphate, tri(β-chloroethyl)phosphate and di(n-butyl) phthalate. The dielectric data are fitted by curves obtained from the Williams and Watt equation (6.31) (solid line), the viscoelastic data by curves obtained from the Davidson–Cole equation (5.14) (dashed line) (from Shears et al. 1974).
+ viscoelastic data
○ dielectric data.

with local reorientational adjustments of molecules. Differences between the dielectric and viscoelastic results exist however, in that the dielectric data are best represented by the Williams–Watts empirical function, equation (6.31) and the viscoelastic retardation data are represented by the Davidson–Cole equation (5.14). A comparison of the characteristic times in these equations, τ_0 and τ_r, is difficult because the dielectric and viscoelastic measurements were made at different frequencies, and hence different temperatures.

Barlow and Erginsav (1974) have been able to compare the viscoelastic retardation time τ_r and the dielectric relaxation time τ_D as a function of temperature for supercooled benzyl benzoate. These results are shown in Fig. 6.24. The high temperature dielectric behaviour, between 85°C and 30°C can be described by a single relaxation time τ_D. At lower temperatures, between −10°C and −43°C, where the liquid is supercooled, the behaviour is described by the Davidson–Cole equation (6.28) with a temperature independent value for β of 0.5. The viscoelastic retardation behaviour in the supercooled region can also be described by the Davidson–Cole equation with $\beta = 0.5$. The retardation time τ_r follows the behaviour observed in other supercooled liquids, being a constant multiple of the Maxwell time τ_m at low temperatures, $\tau_r/\tau_m = (J_r/2J_\infty)^2 = 182$. The dielectric time τ_D is also found to be a constant multiple of τ_m at low temperatures, $\tau_D/\tau_m = 11$. Thus, in this region the dielectric and viscoelastic retardation processes have spectra with the same shape, and the characteristic times τ_D and τ_r have the same temperature dependence as the Maxwell relaxation time τ_m. The rapid variation with temperature of these three times parallels the viscosity variation, and reflects the co-operative nature of molecular reorientation in the supercooled region, the rate at which reorientation can occur being limited by diffusion.

As the temperature is raised to the region of the melting point, around 20°C, molecular motions become less dependent on the cooperation of neighbouring molecules, both τ_r/τ_m and τ_D/τ_m decrease, and the reorientational process is characterised by a single time. The change in β from 0.5 to 1.0 which is observed in the dielectric relaxation is presumably accompanied by a similar change in the retardation behaviour, but experimental limitations prevent this being observed directly.

Also shown in Fig. 6.24 is the single value of a "shear relaxation

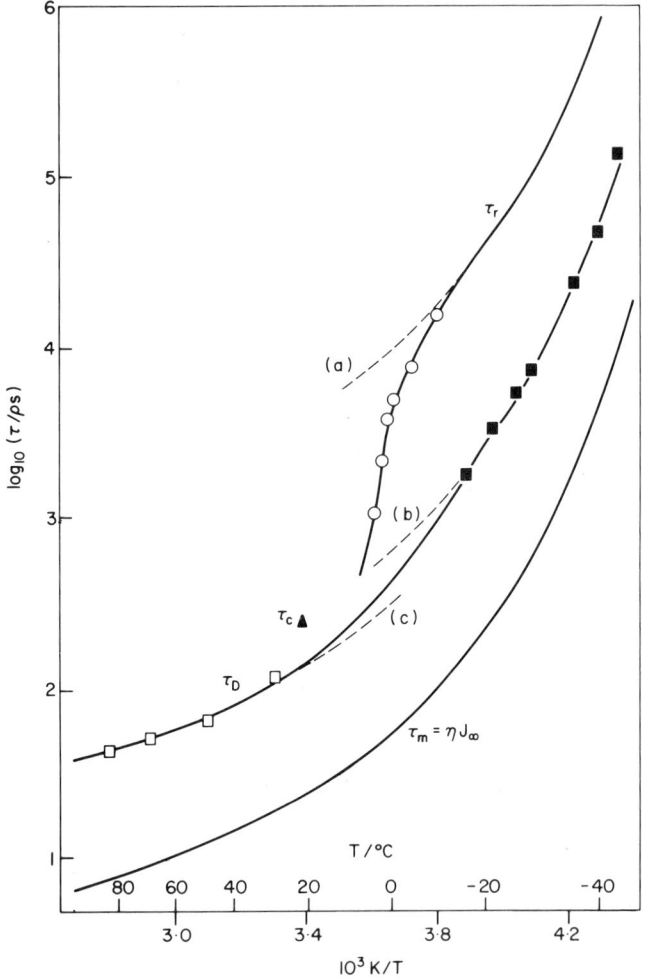

FIG. 6.24. The dielectric relaxation time τ_D, shear retardation time τ_r, Maxwell relaxation time τ_m and the time τ_c from light scattering experiments for benzyl benzoate. Curve (a) is for $\tau_r = 182\,\tau_m$, curve (b) for $\tau_D = 11\,\tau_m$ and curve (c) for $\tau_D = 5.5\,\tau_m$.

○ retardation time τ_r, Davidson–Cole equation (5.14)
■ dielectric relaxation time τ_D, Davidson–Cole equation (6.28)
□ dielectric relaxation time, single relaxation time τ_D
▲ time τ_c from light scattering measurements
(from Barlow and Erginsav, 1974).

time" τ_c as determined from light scattering measurements by Stegeman and Stoicheff (1973) (see Chapter 4, Section 4.5.2). Their results are interpreted in terms of a shear modulus μ_∞ and a time τ_c which are associated with thermally generated shear waves (Rytov, 1958; Leontovich, 1941). The value of τ_c in benzyl benzoate, the only liquid among those studied by light-scattering techniques in which the viscoelastic retardation behaviour could be determined, is clearly much larger than τ_m, τ_D or τ_r. Barlow and Erginsav (1974) suggest that experimental difficulties in the light scattering measurements, at a temperature close to the melting point, could result in the true value of τ_c being somewhat less than the value shown. The value would then become closer to the dielectric time τ_D, and a reasonable extrapolation of τ_r to higher temperature could also lead to a value close to τ_D. It is therefore possible that at temperatures above the melting point τ_c, τ_D and τ_r all have similar values, of the order of 100 ps or less which are associated with the reorientation of individual molecules. The value of μ_∞ found in light-scattering measurements is much closer to the retardational modulus $1/J_r$ than the limiting shear modulus G_∞. This supports recent theories which suggest that the light-scattering behaviour is the consequence of the coupling of molecular reorientations to short range shear and compressional modes rather than shear waves governed by the shear viscosity η and modulus G_∞. A review of the recent literature on this aspect of light-scattering is given by Enright et al. (1972).

The similarities between the dielectric, shear and volume relaxation processes discussed in the preceding paragraphs are not found in all liquids. In many polar molecules, orientation of the dipole can occur without movement of the whole molecule. In such cases, the dielectric and mechanical relaxation times might be expected to be different and to have different temperature dependences. For example, the properties of a molten salt mixture containing calcium and potassium nitrates have been compared by Glover and Matheson (1971). The shear relaxation time, defined as the Maxwell relaxation time $\tau_m = J_\infty \eta_s$, was found to be substantially longer, and much more dependent on temperature, than the mean dielectric relaxation time. Thus the dielectric and shear relaxations would appear to be related to different types of molecular motion. Kono et al. (1966a) have determined the dielectric and mechanical relaxation times in mixtures of glycerol and n-propanol. The ratio of times τ_D/τ_s was found to vary with the concentration of n-propanol, and in pure n-propanol the ratio was markedly dependent on temperature, in contrast to the independence of the ratio with temperature found in pure glycerol. Even though a comparison was made between the dielectric parameters and shear relaxation parameters, instead of the retardation parameters, it appears that here, also, different mechanisms of molecular movement are involved in the dielectric and shear relaxation processes.

It would seem possible, therefore, that in certain cases the study of the dielectric properties of liquids in the supercooled state may provide information which will complement the information obtained from the study of the mechanical properties. If so, the comparative ease with which dielectric measurements can be made would enable liquids to be characterised much more rapidly than by the direct determination of the shear and compressional properties. However, many further measurements of both mechanical and dielectric properties are required, preferably made over the same temperature, pressure and frequency ranges, and on the same samples of liquid, before definite conclusions can be drawn regarding the correlations which might be possible between the properties, and the relationships between the observed properties of the liquid and the fundamental motions of the molecules.

References

Adams, D. R. and Hirst, W. (1973). *Proc. Roy. Soc.* **A332**, 505–525.
Alfrey, T. Jr. (1945). *Quart. Appl. Math.* **3**, 143–150.
Alfrey, T. Jr. and Doty, P. M. (1945). *J. Appl. Phys.* **16**, 700–713.
American Institute of Physics Handbook (1963). 2nd. Ed., Sect. 2n, McGraw-Hill, New York.
American Petroleum Institute (1953). Research Project 44, "Selected values of properties of hydrocarbons and related compounds", Carnegie, Pittsburgh.
American Petroleum Institute (1967). Research Project 42, "Properties of hydrocarbons of high molecular weight", American Petroleum Institute, New York.
Andrade, E. N. da C. (1930). *Nature.* **125**, 309–310.
Andreae, J. H., Bass, R., Heasell, E. L. and Lamb, J. (1958). *Acustica.* **8**, 131–142.
Andreae, J. H. and Joyce, P. L. (1962). *Brit. J. Appl. Phys.* **13**, 462–467.
Andreae, J. H., Edmonds, P. D. and McKellar, J. F. (1965). *Acustica.* **15**, 74–88.
Appledoorn, J. K., Okrent, E. H. and Philippoff, W. (1962). *Proc. Am. Pet. Inst.* **42(III)**, 163–172.
A.S.M.E. Pressure-Viscosity Report (1953). "Viscosity and density of over forty lubricating fluids of known composition at pressures to 150,000 psi and temperatures to 425°F", American Society of Mechanical Engineers, New York.
ASTM (1970). Viscosity-Temperature Charts, Standard D341–43. "Annual Book of ASTM Standards", part 17. American Society for Testing and Materials, Philadelphia.
Barlow, A. J. and Lamb, J. (1959). *Proc. Roy. Soc.* **A253**, 52–69.
Barlow, A. J., Harrison, G., Richter, J., Seguin, H. and Lamb, J. (1961). *Lab. Pract.* **10**, 786–801.
Barlow, A. J., Harrison, G. and Lamb, J. (1964). *Proc. Roy. Soc.* **A282**, 228–251.
Barlow, A. J., Lamb, J. and Matheson, A. J. (1966). *Proc. Roy. Soc.* **A292**, 322–342.
Barlow, A. J. and Subramanian, S. (1966). *Brit. J. Appl. Phys.* **17**, 1201–1214.
Barlow, A. J. and Yazgan, E. (1966). *Brit. J. Appl. Phys.* **17**, 807–819.

Barlow, A. J., Lamb, J., Matheson, A. J., Padmini, P. R. K. L. and Richter, J. (1967a). *Proc. Roy. Soc.* A298, 467–480.
Barlow, A. J., Erginsav, A. and Lamb, J. (1967b). *Proc. Roy. Soc.* A298, 481–494.
Barlow, A. J., Dickie, R. A. and Lamb, J. (1967c). *Proc. Roy. Soc.* A300, 356–362.
Barlow, A. J., Day, M., Harrison, G., Lamb, J. and Subramanian, S. (1969a). *Proc. Roy. Soc.* A309, 497–520.
Barlow, A. J., Erginsav, A. and Lamb, J. (1969b). *Proc. Roy. Soc.* A309, 473–496.
Barlow, A. J., Lamb, J. and Taşköprülü, N. Ş. (1969c). *J. Acoust. Soc. Am.* 46, 569–573.
Barlow, A. J. and Erginsav, A. (1972). *Proc. Roy. Soc.* A327, 175–190.
Barlow, A. J. and Singh, R. P. (1972). *J. Chem. Soc., Farad. Trans. II.* 68, 1404–1409.
Barlow, A. J., Erginsav, A. and Lamb, J. (1972a). *Nature, Physical Science,* 237, 87–88.
Barlow, A. J., Harrison, G., Irving, J. B., Kim, M. G., Lamb, J. and Pursley, W. C. (1972b). *Proc. Roy. Soc.* A327, 403–412.
Barlow, A. J. and Erginsav, A. (1973). *J. Chem. Soc. Farad. Trans. II.* 69, 1200–1207.
Barlow, A. J., Harrison, G., Kim, M. G. and Lamb, J. (1973). *J. Chem. Soc. Farad. Trans. II.* 69, 1446–1453.
Barlow, A. J. and Erginsav, A. (1974). *J. Chem. Soc. Farad. Trans. II.* 70, 885–891.
Barus, C. (1893). *Am. J. Science.* 3rd. series, 45, 87–96.
Bell, J. C., Kannel, J. W. and Allen, C. M. (1964). *J. Basic Eng. Trans. Am. Soc. Mech. Eng.* 86, Series D., 423–432.
Bell, J. F. W., Doyle, B. P. and Smith, B. S. (1966). *J. Sci. Inst.* 43, 28–31.
Berberian, J. G. and Cole, R. H. (1968). *J. Amer. Chem. Soc.* 90, 3100–3104.
Beyer, R. T. and Letcher, S. V. (1969). "Physical Ultrasonics", Academic Press, New York.
Bhatia, A. B. (1967). "Ultrasonic Absorption", University Press, Oxford.
Blair, G. W. Scott (1969). "Elementary Rheology" Chap. 4, Academic Press, London.
Bommel, H. E. and Dransfeld, K. (1960). *Phys. Rev.* 117, 1245–1252.
Born, M. and Wolf, E. (1964). "Principles of Optics" 2nd. Ed. Pergamon Press, London.
Brewer, R. G. and Reickhoff, K. E. (1964). *Phys. Rev. Letters.* 13, 334A–336A.
Bridgman, P. W. (1949). "The physics of high pressure", Bell, London, (reprinted 1971, Dover, New York).
Brillouin, L. (1922). *Ann. Phys.* (France) 9th. ser. 17, 88–122.
Bucaro, J. A. and Carome, E. F. (1967). *J. Acoust. Soc. Am.* 41, 1370–1371.
Buckley, F. and Maryott, A. (1958). "Tables of dielectric dispersion data for pure liquids and dilute solutions". Circular 589, Nat. Bur. Standards, Washington, D.C.

Cady, W. G. (1946). "Piezoelectricity", p. 454, McGraw-Hill, New York.
Capps, W., Macedo, P. B., O'Meara, B. and Litovitz, T. A. (1966). *J. Chem. Phys.* **45**, 3431–3438.
Carpenter, M. R., Davies, D. B. and Matheson, A. J. (1967). *J. Chem. Phys.* **46**, 2451–2454.
Chiao, R. Y. and Stoicheff, B. F. (1964). *J. Opt. Soc. Amer.* **54**, 1286–1287.
Chiao, R. Y., Townes, C. H. and Stoicheff, B. F. (1964). *Phys. Rev. Letters.* **12**, 592–595.
Chu, P. S. Y. and Cameron, A. (1962). *J. Inst. Petroleum.* **48**, 147–155.
Clark, A. E. and Litovitz, T. A. (1960). *J. Acous. Soc. Amer.* **32**, 1221–1236.
Cohen, M. H. and Turnbull, D. (1959). *J. Chem. Phys.* **31**, 1164–1169.
Cole, K. S. and Cole, R. H. (1949). *J. Chem. Phys.* **9**, 341–351.
Cole, R. H. (1960). *Am. Rev. Phys. Chem.* **11**, 149–168.
Crawford, S. M. (1956). *Proc. Phys. Soc. B.* **69**, 1312–1318.
Crook, A. W. (1963). *Phil. Trans.* **A255**, 281–312.
Cummins, H. Z. and Gammon, R. W. (1966). *J. Chem. Phys.* **44**, 2785–2796.
Damon, R. W., Maloney, W. T. and McMahon, D. H. (1970). In "Physical Acoustics" (Mason, W. P. and Thurston, R. N., Eds.) Vol. VII, 273–366, Academic Press, New York.
Dannhauser, W., Child, W. C. Jr. and Ferry, J. D. (1958). *J. Colloid Sci.* **13**, 103–113.
Davidson, D. W. and Cole, R. H. (1951). *J. Chem. Phys.* **19**, 1484–1490.
Davidson, D. W. (1961). *Can. J. Chem.* **39**, 2139–2154.
Davies, D. B. and Matheson, A. J. (1966). *J. Chem. Phys.* **45**, 1000–1006.
Davies, D. B. and Matheson, A. J. (1967). *Trans. Farad. Soc.* **63**, 596–603.
Davies, D. B., Matheson, A. J. and Glover, G. M. (1973). *J. Chem. Soc. Farad. Trans. II.* **69**, 305–314.
Davis, C. M. and Litovitz, T. A. (1965). *J. Chem. Phys.* **42**, 2563–2576.
Debye, P. J. W. (1912). *Ann. Phys.* **39**, 789–839. (trans. in "The collected papers of P. J. W. Debye, 1954", Interscience, N.Y., 650–696.)
Debye, P. J. W. (1913). Bevichte der deutschen physikatischen Gesellschaft, **15**, 777–793 (trans. in "The collected papers of P. J. W. Debye, 1954", Interscience, N.Y., 158–172).
Debye, P. and Sears, F. W. (1932). *Proc. Nat. Acad. Sci.* (USA), **18**, 409–414.
Denney, D. J. (1957). *J. Chem. Phys.* **27**, 259–264.
Dexter, A. R. and Matheson, A. J. (1971). *J. Chem. Phys.* **54**, 3463–3471.
Doolittle, A. K. (1951). *J. Appl. Phys.* **22**, 1471–1475.
Dowson, D. and Higginson, G. R. (1966). "Elasto-hydrodynamic Lubrication", Pergamon Press, Oxford.
Dreisbach, D. (1966). "Liquids and Solutions", Houghton-Mifflin, Boston.
Duncan, W., Hutchins, R. H. and Stewart, P. A. M. (1969). *J. Vac. Sci. Tech.* **6**, 555–558.
Dyson, A. (1965). *Phil. Trans.* **A258**, 529–564.
Dyson, A. (1970). *Phil. Trans.* **A266**, 1–33.
Edmonds, P. D., Pearce, V. F. and Andreae, J. H. (1962). *Brit. J. Appl. Phys.* **13**, 551–560.

Edmonds, P. D. (1966). *Rev. Sci. Inst.* **37**, 367–368.
Eirich, F. R. (Ed.) (1958–70). "Rheology", vols. 1–5, Academic Press, London.
Enright, G. D., Stegeman, G. I. A. and Stoicheff, B. P. (1972). *J. Phys.* (Paris) **33**, C1, 207–213.
Eyring, H. (1936). *J. Chem. Phys.* **4**, 283–291.
Fabelinskii, I. L. (1966). Soviet Physics Uspekhi, **8**, 637–641.
Fabelinskii, I. L. (1968). "Molecular scattering of light", Plenum Press, New York.
Fein, R. S. (1967). *J. Lub. Tech., Trans. Am. Soc. Mech. Eng.* **89**, Series F, 127–131.
Ferry, J. D. (1970). "Viscoelastic properties of polymers", 2nd. Ed. J. Wiley and Sons, Inc., New York.
Fleury, P. A. and Boon, J. P. (1969). *Phys. Rev.* **186**, 244–254.
Fleury, P. A. and Chiao, R. Y. (1966). *J. Acoust. Soc. Am.* **39**, 751–752.
Foster, N. F., Coquin, G. A., Rozgonyi, G. A. and Vannatta, F. A. (1968). *I.E.E.E. Trans.* **SU-15**, 28–41.
Foster, N. F. (1969). *J. Appl. Phys.* **40**, 4202–4204.
Fröhlich, H. (1958). "Theory of Dielectrics", 2nd. Ed., Oxford Univ. Press, Oxford.
Fulcher, G. S. (1925). *J. Amer. Ceram. Soc.* **8**, 339–355.
Glarum, S. H. (1960a). *J. Chem. Phys.* **33**, 639–643.
Glarum, S. H. (1960b). *J. Chem. Phys.* **33**, 1371–1375.
Glover, G. M., Hall, G., Matheson, A. J. and Stretton, J. L. (1968). *J. Phys. E. Sci. Inst.* **1**, 383–388.
Glover, G. M. and Matheson, A. J. (1971). *Trans. Farad. Soc.* **67**, 1960–1970.
Goldblatt, N. R. and Litovitz, T. A. (1967). *J. Acoust. Soc. Am.* **41**, 1301–1307.
Goldstein, M., (1969). *J. Chem. Phys.* **51**, 3728–3739.
Gooberman, G. L. (1969). "Ultrasonics; Theory and Application", English Universities Press, London.
Greenspan, M. (1964). In "Physical Acoustics" (Mason, W. P., Ed.), Vol. IIA, 1–45, Academic Press, New York.
Greenwood, N. N. and Martin, R. L. (1952). *Proc. Roy. Soc.* **A215**, 46–65.
Groenewegen, P. P. M. and Cole, R. H. (1967). *J. Chem. Phys.* **46**, 1069–1074.
Gross, B. (1953). "Mathematical Structure of the Theories of Viscoelasticity", Hermann, Paris.
Gruber, G. J. and Litovitz, T. A. (1964). *J. Chem. Phys.* **40**, 13–26.
Guillemin, E. A. (1947). "Communications Networks", Vol. 2, p. 211, Wiley, New York.
Handbook of Chemistry (1961). (Lange, N. A., Ed.). 10th. Ed. p. 1844, McGraw-Hill, New York.
Harrison, G. and Trachman, E. G. (1972). *J. Lub. Tech., Trans. Am. Soc. Mech. Eng.* **94**, Series E, 306–312.
Hayward, A. T. J. (1967). *Brit. J. Appl. Phys.* **18**, 965–977.
Hemphill, R. B. (1969). *Rev. Sci. Inst.* **40**, 175–176.
Hercher, M. (1968). *Applied Optics*, **7**, 951–966.

Herzfeld, K. F. and Litovitz, T. A. (1959). "Absorption and Dispersion of Ultrasonic Waves", Academic Press, New York.
Higasi, K., Bergmann, K. and Smyth, C. P. (1960). *J. Phys. Chem.* **64**, 880–883.
Hill, N. E., Vaughan, W. E., Price, A. H. and Davies, M. (1969). "Dielectric Properties and Molecular Behaviour", Van Nostrand, London.
Hubbard, J. C. (1931). *Phys. Rev.* **38**, 1011–1019.
Hunston, D. L., Myers, R. R. and Palmer, M. B. (1972a). *Trans. Soc. Rheol.* **16**, 33–44.
Hunston, D. L., Knauss, C. J., Palmer, M. B. and Myers, R. R. (1972b). *Trans. Soc. Rheol.* **16**, 45–57.
Hunt, B. I. and Powles, J. G. (1966). *Proc. Phys. Soc.* **88**, 513–528.
Hutton, J. F. (1968). *Proc. Roy. Soc.* **A304**, 65–80.
Hutton, J. F. and Phillips, M. C. (1969). *J. Chem. Phys.* **51**, 1065–1072.
Hutton, J. F. and Phillips, M. C. (1972). *Nature Physical Science.* **238**, 141–142.
Irving, J. B. and Barlow, A. J. (1971). *J. Phys. E.* **4**, 232–236.
Isakovitch, M. A. and Chaban, I. A. (1966). *Soviet Physics, J.E.T.P.* **23**, 893–905.
Jacquinot, P. (1960). *Repts. Prog. Phys.* **23**, 267–312.
Johari, G. P. and Goldstein, M. (1970). *J. Chem. Phys.* **53**, 2372–2388.
Johnson, K. L. and Cameron, R. (1967). *Proc. Inst. Mech. Eng.* **182**, Pt. 1. 307–319.
Johnson, K. L. and Roberts, A. D. (1974). *Proc. Roy. Soc.* **A337**, 217–242.
Jones, R. V. (1962). *J. Sci. Inst.* **39**, 193–203.
Kauzmann, W. (1948). *Chem. Rev.* **43**, 219–256.
Kim, M. G. (1974). *J. Chem. Soc. Farad. Trans. II.* To be published.
Kinsler, L. E. and Frey, A. R. (1962). "Fundamentals of Acoustics", 2nd. Ed., J. Wiley and Sons, Inc., New York.
Klein, W. R. and Cook, B. D. (1967). *I.E.E.E. Trans. Sonics Ultrasonics,* **SU-14**, 123–134.
Kono, R., Litovitz, T. A. and McDuffie, G. E. Jr. (1966a). *J. Chem. Phys.* **45**, 1790–1796.
Kono, R., McDuffie, G. E. and Litovitz, T. A. (1966b). *J. Chem. Phys.* **44**, 965–970.
Kovacs, A. (1958). *J. Polymer Sci.* **30**, 131–147.
Kumar, S. F. (1963). *Phys. Chem. Glasses.* **4**, 106–111.
Lamb, J. (1965). *In* "Physical Acoustics", (Mason, W. P., Ed.), Vol. IIA, 203–280, Academic Press, New York.
Lamb, J. and Richter, J. (1966). *Electronics Letters.* **2**, 73–74.
Lamb, J. (1967). *Proc. Inst. Mech. Eng.* **182**, Pt. 3A, 293–310.
Landau, L. D. and Lifshitz, E. M. (1969). "Statistical Physics" 2nd. Ed. Addison–Wesley, Reading, Mass.
Laughlin, W. T. and Uhlmann, D. R. (1972). *J. Phys. Chem.* **76**, 2317–2325.
Leaderman, H. (1949). *J. Colloid Sci.* **4**, 193–210.
Lee, R. E. and Booser, E. R. (1970). "Lubrication and Lubricants" in Kirk–Othmer Encyclopedia of Chemical Technology, 2nd. Ed., Vol. **12**, p. 566, Wiley, New York.

Leidecker, H. W. Jr. and La Macchia, J. T. (1968). *J. Acoust. Soc. Amer.* **43**, 143–151.

Leontovich, M. A. (1941). *J. Phys. U.S.S.R.* **4**, 499–514.

Litovitz, T. A. (1952). *J. Chem. Phys.* **20**, 1088–1089.

Litovitz, T. A. and Lyon, T. (1954). *J. Acoust. Soc. Am.* **26**, 577–580.

Litovitz, T. A. and McDuffie, G. E. Jr. (1963). *J. Chem. Phys.* **39**, 729–734.

Litovitz, T. A. and Davies, C. M. (1965). *In* "Physical Acoustics" (Mason, W. P., Ed.), Vol. IIA, 281–349, Academic Press, New York.

Lucas, R. and Biquard, P. (1932). *J. Phys. Rad.* 7th ser. **3**, 464–477.

McCrum, N. G., Read, B. E. and Williams, G. (1967). "Anelastic and dielectric effects in polymeric solids", Wiley, London.

Macedo, P. B. and Litovitz, T. A. (1965). *J. Chem. Phys.* **42**, 245–256.

Macedo, P. B., Simmons, J. H. and Maller, W. (1968). *Phys. & Chem. of Glasses*, **9**, 156–164.

Mackenzie, J. D. and Murphy, W. K. (1960). *J. Chem. Phys.* **33**, 366–369.

McSkimin, H. J. (1952). *J. Acoust. Soc. Am.* **24**, 355–365.

McSkimin, H. J. (1961). *J. Acoust. Soc. Am.* **33**, 12–16.

McSkimin, H. J. (1962). *J. Acoust. Soc. Am.* **34**, 404–409.

McSkimin, H. J. (1965). *J. Acoust. Soc. Am.* **37**, 864–871.

McSkimin, H. J. and Andreatch, P. Jr. (1967a). *J. Acoust. Soc. Am.* **42**, 248–252.

McSkimin, H. J. and Andreatch, P. Jr. (1967b). **41**, 1052–1057.

McSkimin, H. J. and Bateman, T. B. (1969). *J. Acoust. Soc. Am.* **45**, 852–858.

McSkimin, H. J. (1970). *J. Acoust. Soc. Am.* **47**, 163–167.

Magill, J. H. and Li, H-M. (1973). *Journal of Polymer Sci. (B), Polymer Letters Edition.* **11**, 667–672.

Mason, W. P. (1947). *Trans. Am. Soc. Mech. Eng.* **69**, 359–367.

Mason, W. P., Baker, W. O., McSkimin, H. J. and Heiss, J. H. (1949). *Phys. Rev.* **75**, 936–946.

Mason, W. P. (1950). "Piezoelectric crystals and their applications to ultrasonics", Van Nostrand, Princeton, New Jersey.

Matheson, A. J. (1971a). "Molecular Acoustics", Wiley, London.

Matheson, A. J. (1971b). *J. Phys. E., Sci. Inst.* **4**, 796.

Matsumoto, A. and Higasi, K. (1962). *J. Chem. Phys.* **36**, 1776–1780.

Maxwell, J. C. (1868). *Phil. Mag.* 4th. series, **35**, 129–145.

May, J. E. Jr. (1964). *In* "Physical Acoustics" (Mason, W. P., Ed.), Vol. 1A, 417–483, Academic Press, New York.

Mayer, W. G. (1964). *J. Acoust. Soc. Am.* **36**, 779–781.

Meister, R., Marhoeffer, C. J., Sciamanda, R., Cotter, L. and Litovitz, T. A. (1960). *J. Appl. Phys.* **31**, 854–870.

Mielenz, K. D., Stephens, R. B. and Nefflen, K. F. (1964). *J. Res. Nat. Bureau Standards.* **68C**, 1–6.

Montrose, C. J., Solovyev, V. A. and Litovitz, T. A. (1968). *J. Acoust. Soc. Am.* **43**, 117–130.

Montrose, C. J. and Litovitz, T. A. (1970). *J. Acoust. Soc. Amer.* **47**, 1250–1257.

Moore, R. S., McSkimin, H. J., Gieniewski, C. and Andreatch, P. (1969). *J. Chem. Phys.* **50**, 466–472.
Mopsik, F. I. and Cole, R. H. (1966). *J. Chem. Phys.* **44**, 1015–1019.
Moreno, T. (1958). "Microwave Transmission Design Data", p. 141, Dover, New York.
Mountain, R. D. (1966). *J. Res. Nat. Bur. Stands.* **70A**, 207–220.
Newton, I. (1687). "Principia", translation in Reiner, M. (1960). "Deformation, Strain and Flow", Lewis, London.
Nolle, A. W. (1950). *J. Polymer Sci.* **5**, 1–54.
Nozdrev, V. F. (1965). "Use of Ultrasonics in Molecular Physics", Pergamon, Oxford.
O'Connor, C. L. and Schluff, J. P. (1966). *J. Acoust. Soc. Am.* **40**, 663–666.
Oldroyd, J. G. (1956). *In* "Rheology" (Eirich, F. R., Ed.), Vol. 1, 653–682, Academic Press, New York.
O'Neill, H. T. (1949). *Phys. Rev.* **75**, 928–935.
Papadakis, E. P. (1967). *J. Acoust. Soc. Am.* **42**, 1045–1051.
Paul, G. R. and Cameron, A. (1974). *Nature.* **248**, 219–220.
Pellam, J. R. and Galt, J. K. (1946). *J. Chem. Phys.* **14**, 608–614.
Philippoff, W. (1963). *J. Appl. Phys.* **34**, 1507–1511.
Philippoff, W. (1964). *Trans. Soc. Rheol.* **8**, 117–135.
Phillips, M. C., Barlow, A. J. and Lamb, J. (1972). *Proc. Roy. Soc.* **A329**, 193–218.
Piccirelli, R. and Litovitz, T. A. (1957). *J. Acoust. Soc. Am.* **29**, 1009–1020.
Piercy, J. E. and Rao, M. G. S. (1967). *J. Acoust. Soc. Am.* **41**, 1063–1073.
Pinkerton, J. M. M. (1947). *Nature.* **160**, 128–129.
Pinnow, D. A., Candau, S. J., La Macchia, T. J. and Litovitz, T. A. (1968a). *J. Acoust. Soc. Am.* **43**, 131–142.
Pinnow, D. A., Candau, S. J. and Litovitz, T. A. (1968b). *J. Chem. Phys.* **49**, 347–362.
Plazek, D. J. and Magill, J. H. (1966). *J. Chem. Phys.* **45**, 3038–3050.
Powles, J. G. (1953). *J. Chem. Phys.* **21**, 633–637.
Pryde, J. A. (1966). "The Liquid State", Hutchinson, London.
Quate, C. F., Wilkinson, C. D. W. and Winslow, D. K. (1965). *Proc. I.E.E.E.* **53**, 1604–1623.
Raman, C. V. and Nath, N. S. N. (1935). *Proc. Ind. Acad. Sci.* **A2**, 406–420.
Raman, C. V. and Nath, N. S. N. (1936). *Proc. Ind. Acad. Sci.* **A3**, 75–84.
Rank, D. H., Hollinger, A. and Eastman, D. P. (1966). *J. Opt. Soc. Am.* **56**, 1057–1058.
Reiner, M. (1958). *In* "Handbuck der Physik" (Flügge, S., Ed.), Vol. VI, 434–550, Springer-Verlag, Berlin.
Reiner, M. (1960). "Deformation, Strain and Flow", Lewis, London.
Roelands, C. J. A. (1966). "Correlational aspects of the viscosity–temperature–pressure relationship of lubricating oils". Doctoral thesis, Univ. of Delft, Netherlands.
Roscoe, R. (1950). *Brit. J. Appl. Phys.* **1**, 171–173.

Rouse, P. E. Jr., Bailey, E. D. and Minkin, J. A. (1950). *Proc. Am. Pet. Inst.* **30 III**, 54—78.
Rowlinson, J. S. (1969). "Liquids and liquid mixtures", Butterworth, London.
Rytov, S. M. (1958). *Sov. Phys. J.E.T.P.* **6**, 401—408 and 513—523.
Schaafs, W. (1967). *In* "Landolt—Börnstein: Numerical Data and Functional Relationships in Science and Technology", New Series, Group II, Vol. **5**, (Molecular Acoustics)(Hellwege, K. M. and Hellwege, A. M., Eds.), Springer-Verlag, Berlin.
Schwarzl, F. and Struik, L. C. E. (1967). *Adv. Mol. Relax. Processes.* **1**, 201—255.
Schwarzl, F. (1969). *Rheol. Acta.* **8**, 6—17.
Seki, H., Granato, A. and Truell, R. (1956). *J. Acoust. Soc. Am.* **28**, 230—238.
Shapiro, S. L. and Broida, H. P. (1967). *Phys. Rev.* **154**, 129—138.
Shears, M. F., Williams, G., Barlow, A. J. and Lamb, J. (1974). *J. Chem. Soc. Trans. Faraday II.* **70**, 1783—1793.
Simmons, J. H. and Macedo, P. B. (1968). *J. Acoust. Soc. Am.* **43**, 1295—1301.
Slie, W. M., Donfor, A. R. and Litovitz, T. A. (1966). *J. Chem. Phys.* **44**, 3712—3718.
Slie, W. M. and Madigosky, W. M. (1968). *J. Chem. Phys.* **48**, 2810—2817.
Smith, F. W. (1960). *Trans. Amer. Soc. Lub. Engrs.* **3**, 18—25.
Smith, R. A. (1970). *Endeavour.* **29**, 71—76.
Smyth, C. P. (1955). "Dielectric behaviour and Structure", McGraw-Hill, New York.
Smyth, C. P. (1966). *Rev. Phys. Chem.* **17**, 433—456.
Solovyev, V. A., Montrose, C. J., Watkins, M. H. and Litovitz, T. A. (1968). *J. Chem. Phys.* **48**, 2155—2162.
Starunov, V. S., Tiganov, E. V. and Fabelinskii, I. L. (1966). *J.E.T.P. Letters.* **4**, 176-179.
Starunov, V. S., Tiganov, E. V. and Fabelinskii, I. L. (1967). *J.E.T.P. Letters.* **5**, 260—262.
Stearns, R. S. (1970). "Viscometry" in Kirk—Othmer Encyclopedia of Chemical Technology, 2nd. Ed., Vol. **21**, 473—474, Wiley, New York.
Stegeman, G. I. A. and Stoicheff, B. P. (1968). *Phys. Rev. Letters.* **21**, 202—206.
Stegeman, G. I. A. and Stoicheff, B. P. (1973). *Phys. Rev.* 3rd Series A. **7**, 1160—1177.
Stewart, E. S. and Stewart, J. L. (1963). *J. Acoust. Soc. Am.* **35**, 975—981.
Stratton, R. A. (1966). *J. Colloid Sci.* **22**, 517—530.
Tammann, G. and Hesse, W. (1926). *Z. anorg. allgem. Chem.* **156**, 245—257.
Taşköprülü, N. Ş., Barlow, A. J. and Lamb, J. (1961). *J. Acoust. Soc. Am.* **33**, 278—285.
Tauke, J., Litovitz, T. A. and Macedo, P. B. (1968). *J. Am. Ceram. Soc.* **51**, 158—163.
Thurston, R. N. (1964). *In* "Physical Acoustics" (Mason, W. P., Ed.), Vol. 1A, pp. 1—110, Academic Press, New York.
Timmermans, J. (1965). "Physico-chemical constants of pure organic compounds", 2nd. Ed., Elsevier, New York.

Tittmann, B. R. and Bommel, H. E. (1967). *Rev. Sci. Inst.* **38**, 1491–1496.
Tobolsky, A. V. (1958). *In* "Rheology" (Eirich, F. R., Ed.), Vol. II, 63–81, Academic Press, New York.
Truesdell, C. (1953). *J. Rat. Mech. Anal.* **2**, 594–616.
Vaughan, W. E., Lovell, W. S. and Smyth, C. P. (1962). *J. Chem. Phys.* **36**, 753–758.
Wagner, K. W. (1913). *Ann. der Physik.* **40**, 817–855.
Walther, C. (1931). *Erdol und Teer.* **7**, 382.
Whitehead, S. (1944). *J. Sci. Inst.* **21**, 73–80.
Willard, G. W. (1949). *J. Acoust. Soc. Am.* **21**, 101–108.
Williams, M. L., Landel, R. F. and Ferry, J. D. (1955). *J. Amer. Chem. Soc.* **77**, 3701–3707.
Williams, G. and Watts, D. C. (1970). *Trans. Farad. Soc.* **66**, 80–85.
Williams, G. and Hains, J. H. (1973). *J. Chem. Soc.* Faraday Symposium, **6**, 14–22.
Winslow, J. W., Good, R. J. and Berghausen, P. E. (1957). *J. Chem. Phys.* **27**, 309–312.
Worster, R. C. (1951). *Proc. Inst. Mech. Eng.* **165**, 269–270.
Yager, W. A. (1936). *Physics* (later *J. Appl. Phys.*) **7**, 434–450.
Yazgan, E. (1966). Ph.D. Thesis, University of Glasgow.

Index

Absorption
    of longitudinal waves, 65, 91, 152, 154–156
    of shear waves, 34–5, 64
Arrhenius,
    behaviour, 130–131
    equation, 20
    temperature, 25

B.E.L. equation, 117–121
Bonding of transducers, 78–79, 93
Bragg diffraction of light, 98–99
Brillouin scattering, 100–106
    stimulated, 106–107
Bulk modulus, 13, 154–156
    secant, 13, 15
Bulk viscosity, 152, 155
Burgers model system, 47

Compliance, complex, 37
    high frequency, 38, 114
    retardational, 133–136
Compressibility, 13
Cooperative behaviour, 130–131, 178
Creep response,
    Burgers model, 48
    Maxwell element, 43
    Voigt element, 46

Davidson–Cole equation, 133, 176
Davidson–Cole distribution, 128
Debye process, 174–175
Density of liquids, 12–15
    temperature variation, 13
    pressure variation, 13

Dielectric relaxation, 173–186
Diffraction,
    of light by longitudinal waves, 96–100
    of longitudinal waves, 93
    of shear waves, 77
Diffusion of molecules, 26, 147–149, 178
Distribution,
    of relaxation times, 52, 120, 128
    of retardation times, 53, 133
Dynamic viscosity, 37, 41, 119

Elasticoviscous behaviour, 32
Elastohydrodynamic lubrication, 6–8, 172–173
Equilibrium compliance, 38, 133, 139–140
Experimental techniques,
    high pressure, 141
    longitudinal wave, 91–95
    optical, 96–107
    shear wave, 65–91

Fabry–Perot interferometer, 101
Free volume, 21, 26, 130, 139, 147
    equation, 21
Fluidity, 38

Gaussian distribution, 128, 144
Glass transition temperature, 2, 22, 64, 111, 114
Glassy state, 12, 22

INDEX

High pressure technique, 141
Hooke's Law, 31

Impedance, mechanical, 35, 38, 69, 94
Inclined incidence technique, 72–73

Light scattering, 96–107
Liquid state, 11–12
Liquids, supercooled, 12, 23
Longitudinal waves,
    absorption, 65, 91, 152, 154–156
    propagation, 153–156
    velocity, 93, 95, 154
Lubrication, elastohydrodynamic,
    6–8, 172–173

Magnetostriction, 87
Master curve, 56
Maxwell element, 39–43, 118
Maxwell relaxation time, 3, 40, 63,
    133, 139, 183, 184
Mechanical impedance, 35, 38, 69, 94
Method of reduced variables, 56–61,
    109, 115, 149
Modulus,
    bulk, 13, 154–156
    complex shear, 35–38
    limiting high frequency shear, 4,
        23, 37, 110–116, 129, 141–143
    longitudinal, 153, 156–8
    secant bulk, 13, 15
Molecular diffusion, 26, 147–149, 178
Molecular reorientation, 181–3
Molecular rotation, 26, 138–9

Neper, 92, 159
Newtonian liquid, 19
Newton's Law, 15, 31
Non-Newtonian behaviour, 19
Normal incidence technique, 69–72
Normalised variables, 61

Optical techniques, 96–107

Pressure,
    effect on density, 13
    effect on limiting high frequency
        modulus, 141–144
    effect on viscoelastic behaviour,
        144–145
    effect on viscosity, 28–30
Pulse technique,
    for longitudinal waves, 91–95
    for shear waves, 69–84
    superposition methods, 81–84

Quartz transducers,
    longitudinal, 91
    shear, 78
    torsional, 84

Raman–Nath scattering of light, 97–9
Rayleigh scattering of light, 102
Rayleigh wing, 104
Reduced variables, 56–61
Relaxation time, Maxwell, 3, 40, 63,
    133, 139, 183, 184
Relaxation times, spectrum of, 52, 60,
    120, 128
Reorientation of molecules, 138–139
Resonance technique, 90–91
Retardation time, Voigt, 45
Retardation times, spectrum of, 53,
    61, 133
Retardational compliance, 133–136
Rotation, molecular, 26, 139

Scattering of light,
    by longitudinal waves, 96–107
    by shear waves, 105–6, 185
Secant bulk modulus, 13, 15
Shear mechanical impedance, 35, 38,
    69
Shear waves,
    in liquids, 32–38
    in solids, 70–72, 88
    reflection of, 66–69, 72
Skewed-arc function, 176–7

INDEX

Spectrum,
    of relaxation times, 52, 60, 120, 128
    of retardation times, 53, 61, 133
Stimulated Brillouin scattering, 106–107
Stress relaxation, 2, 42
Structural
    relaxation, 151–153
    analysis of data, 156–163
    in liquids, 163–173
Structural retardation, 172
Supercooled liquid, 12, 23

Theories,
    of liquid state, 5
    of viscoelastic behaviour, 145–149
Time–temperature superposition, 56–61, 109, 115, 149
Torsional transducers, 84, 87
Torsional waves, 84–88
Transducers,
    longitudinal, 91
    shear, 78
    torsional, 84, 87
Travelling wave techniques, 84–90
Transient response, 2, 41, 46, 48

Ultrasonic waves *see* longitudinal waves

Velocity of longitudinal waves, 93, 95, 153–156, 159–160
Velocity of shear waves, 34
V.H.F. techniques, 80
Viscoelasticity,
    theory of, 31–61
    mathematical aspects, 32, 56, 57
Viscoelastic behaviour, 32
Viscoelastic relaxation,
    effect of pressure, 144–145
    in liquids, 116–132
    in mixtures, 121–128
    theories of, 145–149
Viscoelastic retardation in liquids, 132–141
Viscosity,
    bulk, 152
    dynamic, 37, 41, 119
    equations, 20–30
    pressure variation, 28–30
    temperature variation, 20
    units, 17
    values, of liquids, 15–19
    volume, 152, 155
Voigt element, 43–46